BIOTECHNOLOGY IN AGRICULTURE,
INDUSTRY AND MEDICINE

CHITOSAN NANOPARTICLES FOR BIOMEDICAL APPLICATIONS

BIOTECHNOLOGY IN AGRICULTURE, INDUSTRY AND MEDICINE

Additional books in this series can be found on Nova's website under the Series tab.

Additional E-books in this series can be found on Nova's website under the E-book tab.

NANOTECHNOLOGY SCIENCE AND TECHNOLOGY

Additional books in this series can be found on Nova's website under the Series tab.

Additional E-books in this series can be found on Nova's website under the E-book tab.

BIOTECHNOLOGY IN AGRICULTURE,
INDUSTRY AND MEDICINE

CHITOSAN NANOPARTICLES FOR BIOMEDICAL APPLICATIONS

PAULA PEREIRA
VERA CARVALHO
REINALDO RAMOS
AND
MIGUEL GAMA

Nova Biomedical Books
New York

Copyright © 2010 by Nova Science Publishers, Inc.

All rights reserved. No part of this book may be reproduced, stored in a retrieval system or transmitted in any form or by any means: electronic, electrostatic, magnetic, tape, mechanical photocopying, recording or otherwise without the written permission of the Publisher.

For permission to use material from this book please contact us:
Telephone 631-231-7269; Fax 631-231-8175
Web Site: http://www.novapublishers.com

NOTICE TO THE READER

The Publisher has taken reasonable care in the preparation of this book, but makes no expressed or implied warranty of any kind and assumes no responsibility for any errors or omissions. No liability is assumed for incidental or consequential damages in connection with or arising out of information contained in this book. The Publisher shall not be liable for any special, consequential, or exemplary damages resulting, in whole or in part, from the readers' use of, or reliance upon, this material. Any parts of this book based on government reports are so indicated and copyright is claimed for those parts to the extent applicable to compilations of such works.

Independent verification should be sought for any data, advice or recommendations contained in this book. In addition, no responsibility is assumed by the publisher for any injury and/or damage to persons or property arising from any methods, products, instructions, ideas or otherwise contained in this publication.

This publication is designed to provide accurate and authoritative information with regard to the subject matter covered herein. It is sold with the clear understanding that the Publisher is not engaged in rendering legal or any other professional services. If legal or any other expert assistance is required, the services of a competent person should be sought. FROM A DECLARATION OF PARTICIPANTS JOINTLY ADOPTED BY A COMMITTEE OF THE AMERICAN BAR ASSOCIATION AND A COMMITTEE OF PUBLISHERS.

Additional color graphics may be available in the e-book version of this book.

Library of Congress Cataloging-in-Publication Data

Chitosan nanoparticles for biomedical applications / Paula Pereira ... [et al.].
      p. ; cm.
  Includes bibliographical references and index.
  ISBN 978-1-61761-098-1 (softcover)
  1. Chitosan--Biotechnology. 2. Nanoparticles. I. Pereira, Paula.
  [DNLM: 1. Chitosan--therapeutic use. 2. Drug Delivery Systems. 3. Nanoparticles--therapeutic use. QU 83]
  TP248.65.C55C556 2010
  610.28--dc22
                           2010029768

Published by Nova Science Publishers, Inc. † New York

Contents

Preface		vii
Chapter I	Introduction	1
Chapter II	Production of Chitosan-Based Nanoparticles	3
Chapter III	Biomedical Applications	11
Chapter IV	Conclusions	47
Acknowledgments		49
References		51
Index		71

Preface

Chitosan is a rather abundant material with exquisite properties, which may be processed into a variety of materials including hydrogels, fibres, membranes, etc. The production of chitosan-based nanogels, also known as macromolecular miceles, has been successfully achieved using different techniques, which will be reviewed. This book covers the properties and applications of chitosan nanogels in the biomedical field, namely as a drug delivery vehicle for biopharmaceuticals. The main achievements and recent developments will be addressed.

Chapter I

Introduction

Chitosan (poly[β-(1-4)-2-amino-2-deoxy-D-glucopyranose]) is a natural, non-toxic and biodegradable linear polysaccharide, composed by β-(1-4)-linked *N*-glucosamine and *N*-acetyl-glucosamine residues [1]. Chitosan (CS) is obtained upon partial alkaline deacetylation of chitin, a structural element in the exoskeleton of crustaceans and insects, which is the second-most abundant polysaccharide next to cellulose [2]. Nevertheless, being insoluble in water and chemically inert, applications of chitin are limited. In turn, having hydroxyl and amine reactive groups, CS is susceptible to structural modifications.

CS is insoluble in water and organic solvents, but soluble in dilute aqueous acidic solutions (pH < 6.5), due to the protonation of the glucosamine residues into the soluble form $R-NH_3^+$ [3]. Commercial CS has an average molecular weight (Mw) ranging from 3800 to 2 000 000 Daltons (Da) and a deacetylation degree - the proportion of glucosamine residues - of 66 to 100% [4]. The deacetylation degree (DD) has an important influence on various properties including solubility, biodegradability, toxicity, antimicrobial activity, etc. [5]. Furthermore, CS is non-toxic, stable, biodegradable and sterilizable. It is mucoadhesive, due to electrostatic interactions with mucosal surfaces [6, 7]. The reactivity and polycationic character allow the production of a variety of formulations with different properties, ranging from hydrogels, rods and fibres to nano/microparticles and membranes [8]. Altogether, these properties make of medical grade CS a versatile material with extensive application in the biomedical and biotechnological fields [9]. Table 1 summarizes some relevant biomedical applications of CS.

Table 1. Biomedical applications of CS

Biomedical application	Remarks	Ref.
Artificial skin and wound healing	Structural similarity with glycosaminoglycans makes CS and its derivatives eligible for skin replacement. CS and its derivatives support blood coagulation, prevent abnormal fibroblastic reactivity, and act as a bactericide and wound-healing accelerators.	[10-13]
Ophthalmology	CS possesses all the characteristics required for making an ideal contact lens: optical clarity, mechanical stability, satisfactory optical correction, gas permeability and wettability.	[14]
Blood anticoagulants	Sulfonated derivatives act as blood-thinner and lipoprotein lipase (LPL)-releasing agents	[15]
Orthopedic/periodontal applications	Calcium-based compounds with chitin and CS as bone substitutes	[16-19]
Tissue engineering	Porous biodegradable matrixes for cell seeding	[19-23]
Antimicrobial applications	Inhibition of microorganisms growth	[24, 25]

One of the most well-known biomedical applications is the development of drug delivery systems. In fact, papers related to drug delivery systems using CS are available by the hundreds [26-39]. Different types of CS-made drug carriers have been conceived for various administration routes, such as oral, sub-lingual, nasal, transdermal, parenteral, vaginal, cervical, intrauterine and rectal.

Chapter II

Production of Chitosan-Based Nanoparticles

CS microspheres and nanoparticles (NPs) are widely studied drug delivery systems. The use of NPs offers many advantages, providing targeted delivery of drugs, improving the bioavailability and stability of the therapeutic agents against chemical/enzymatic degradation [40]. Micro/nanoparticles may hold drug molecules attached to the matrix, dissolved, encapsulated or entrapped [41].

Different methods have been used to prepare CS particulate systems. The selection of one of these methods depends upon the specificities and requirements associated to each application, namely the physicochemical properties and thermal-chemical stability of the active agent, the envisaged release kinetic profile, biodistribution, cellular uptake efficiency and intracellular fate of the NPs, etc.

Emulsion Crosslinking

This technique is based on the reaction between the primary amines and a multifunctional crosslinking agent bearing aldehyde groups. In this process, a CS solution in acetic acid is emulsified in liquid paraffin (w/o emulsion). The aqueous droplets are stabilized using a suitable surfactant. The emulsion is then reticulated with an appropriate crosslinking agent such as glutaraldehyde, to stabilize the polysaccharide droplets (Figure 1). The amount of crosslinking agent varies according to the crosslinking density required. The nanospheres are then washed and dried [42-44]. The incorporation of drugs is achieved by dispersion in the CS solution, in the beginning of the process, becoming entrapped during the crosslinking reaction. The entrapment efficiency may be improved by performing multiple emulsions [45]. Major drawbacks of this method are associated with the use of organic solvents and crosslinking agents, that may adversely affect the stability of proteins [46]. The complete removal of the unreacted - often toxic - crosslinker is difficult to achieve. Moreover, glutaraldehyde crosslinked NPs present negative effects on cell viability.

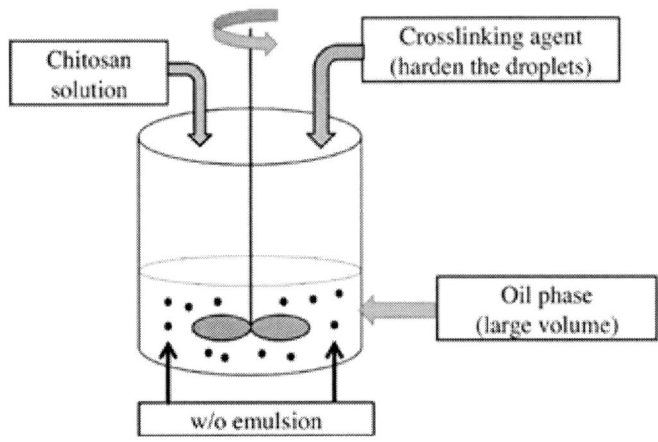

Figure 1. Emulsion crosslinking methodology for the preparation of CS NPs.

Coacervation/Precipitation

This method takes advantage of the physicochemical properties of CS, specifically its insolubility in alkaline pH. CS is dissolved in acidic solution and precipitates/coacervates in contact with an alkaline solution. Spraying the CS solution into sodium hydroxide, NaOH-methanol or ethanediamine alkaline solutions, using compressed air, originates coacervated droplets, forming the NPs [47]. Separation and purification of the particles is finally achieved by centrifugation, followed by successive washing steps with hot and cold water. The size of the NPs can be controlled by varying the diameter of the compressed air nozzle. Berthold et al. [48] described a different approach: the formation of the particles is obtained after addition of sodium sulfate solution to a CS solution, under mild agitation and continuous sonication for 30 min. However, a crosslinking agent is frequently used to harden the NPs.

Emulsion-Droplet Coalescence Method

This technique was developed by Tokumitsu *et al.* [49]. In this approach, precipitation is induced by allowing CS droplets to combine with NaOH droplets. This method involves both emulsion crosslinking and precipitation. A stable emulsion containing the aqueous CS solution along with the drug in liquid paraffin oil is produced; a second emulsion, containing a NaOH solution, is produced in a similar way. When both emulsions are mixed under high-speed stir, droplets of each emulsion collide at random, coalesce, and finally precipitate as small size particles. NPs are obtained within the emulsion-droplets. The method is schematically shown in Figure 2. The particle size varies inversely with the DD of CS, the larger NPs containing lower amounts of drug.[49]

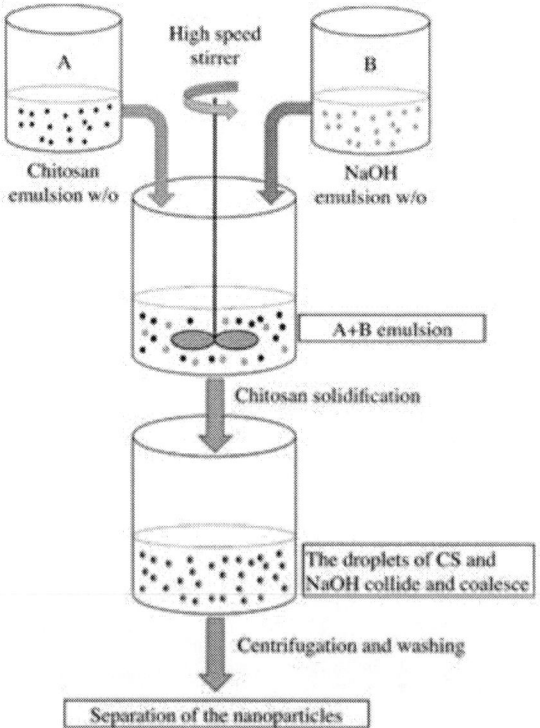

Figure 2. Production of CS NPs using the emulsion-droplet coalescence technique.

Ionotropic Gelation

This method is based on the electrostatic interactions between the CS amine group and a polyanion such as tripolyphosphate (TPP) [50-52]. TPP, a non-toxic multivalent anion, is often used to prepare CS NPs [53]. The complexation between oppositely charged macromolecules as a way to prepare CS NPs has attracted much attention, because the process is very simple and mild: the acidic solution of CS is added dropwise, under constant stirring, to the polyanionic TPP solution. Due to the interaction between oppositely charged species, ionic gelation occurs, giving rise to spherical particles (Figure 3). The particles size and surface charge can be modified by varying the ratio of CS and stabilizer. The efficiency of the method is dependent upon the deacetylation of CS. In fact, the gelation process occurs through the interaction of the protonated amino groups of CS. One of the major drawbacks of this technique is the poor stability and mechanical properties of the NPs, thus limiting their usage in drug delivery. Furthermore, the separation and redispersion processes are difficult to achieve. The negatively charged DNA can also form polyelectrolyte complexes with cationically charged CS, through ionic gelation, as demonstrated by Mansouri *et al.* [54]. The CS properties, namely the Mw, DD and concentration used, have a significant impact on those of the NPs and on its performance as drug carriers [55-57].

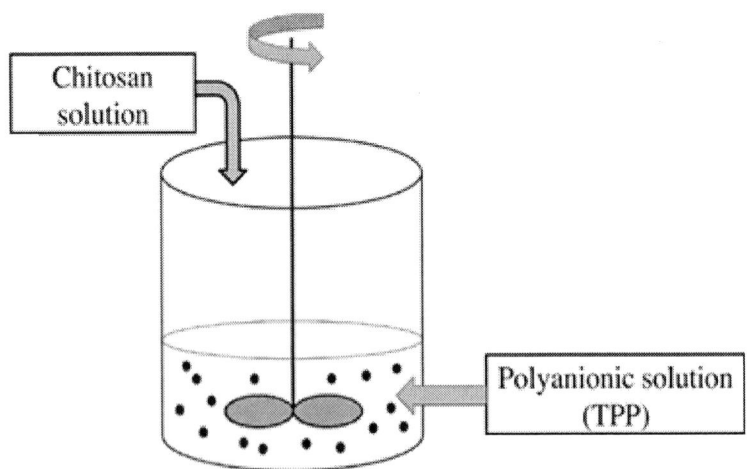

Figure 3. Preparation of CS NPs by the ionotropic gelation method.

Reverse Micelles Method

Reverse micelles are thermodynamically stable liquid mixtures of water, oil and surfactant. Macroscopically, the emulsion is homogeneous and isotropic, structured on a microscopic scale, with the aqueous and oil microdomains separated by surfactant-rich films. The aqueous core of the reverse micellar droplets can be used as a nanoreactor to prepare NPs. Since the size of the obtained reverse micellar droplets usually lies between 1 and 10 nm [58], the preparation of drug-loaded NPs will produce extremely fine particles with a narrow size distribution. In this technique, the reverse micelles are formed by dissolving a surfactant into an organic solvent, giving rise to a water-in-oil micellar system (Figure 4). The aqueous phase containing the CS and the drug are added to this emulsion with constant vortexing and the NPs forms in the core of the reverse micelles. An additional amount of water may be added to produce larger sized NPs. The maximum amount of drug that can be dissolved in reverse micelles has to be determined for each case, by gradually increasing the amount of drug, until the clear microemulsion is transformed into a translucent solution. To this transparent solution, a cross-linking agent is added with constant overnight stirring. The organic solvent is then evaporated. The material is redispersed in water with sonication and the surfactant is salted out. The mixture is finally centrifuged and the supernatant solution, which contains the drug-loaded NPs, is dialyzed for about 1 h and lyophilized to dry powder [59].

Template Polymerization

In this technique, the NPs are obtained upon template polymerization of an acrylic monomer next to the chitosan backbone. CS is firstly dissolved in an acrylic monomer solution under magnetic stirring. Due to the electrostatic interaction, the negatively charged acrylic monomers align along the chitosan molecules. After complete dissolution of CS, the polymerization is started by adding the initiator ($K_2S_2O_8$) under stirring at 70°C. The complete polymerization leads to the appearance of an opalescent solution, indicating the NPs formation. The NPs solution are then filtered and dialysed to remove the residual monomers and initiator. The obtained NPs are positively charged and present a size in the range of 50 to 400 nm [60, 61].

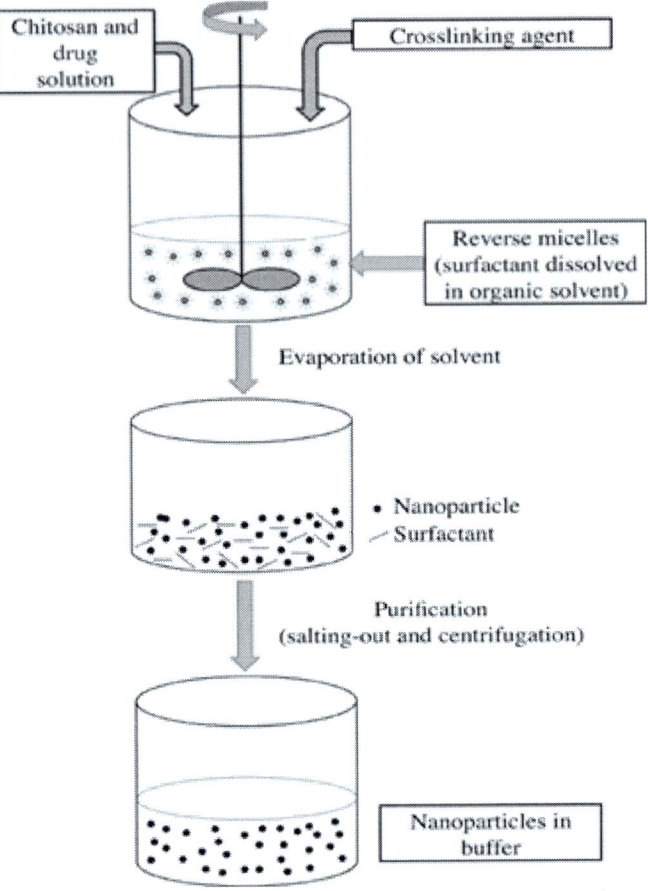

Figure 4. Reverse micelles method of preparation of CS NPs.

Self-Assembled Nanogels

The self-assembly process, defined as the autonomous organization of components into structurally well-defined aggregates, is characterized by numerous beneficial attributes; it is cost-effective, versatile and facile; the process occurs towards the system's thermodynamic minima, resulting in stable and robust structures. Molecular self-assembly is characterized by diffusion followed by specific association of molecules through non-covalent interactions, including electrostatic and/or hydrophobic associations. Individually, such interactions are weak, but dominate the structural and conformational behavior of the assembly due to the large number of interactions involved. While oppositely charged polysaccharides associate readily as a result of electrostatic attractions, interactions among neutral polysaccharides tend to be weaker, or nonexistent, a modification with chemical entities able to trigger assembly being necessary [62]. Hydrophobically modified chitosan is an interesting strategy, consisting on chemical linkage of hydrophobic grafts on hydrophilic backbone, to induce the formation of NPs via hydrophobic interactions [63-67]. Upon contact with an aqueous environment, hydrophobically modified chitosan spontaneously form self-aggregated NPs, via intra- or intermolecular associations between the hydrophobic moieties, primarily to minimize the interfacial free energy.

Chapter III

Biomedical Applications

The biomedical application for CS NPs is particularly relevant concerning the development of delivery systems for biopharmaceuticals, although many papers describe also its use as carriers for low molecular weight drugs. The recent progresses in these biomedical applications are reviewed in this section.

3.1. Protein, Peptide and Oligosaccharide Delivery

Therapeutic proteins are becoming available for the treatment of a wide range of diseases, such as cancer, autoimmune diseases and metabolic disorders. A main problem limiting the efficiency of protein therapeutics is the reduced stability and short circulation half-lives after parenteral administration (i.e. intravenous, intramuscular, or subcutaneous) [68]. As a result of the invasive nature, injectable formulations are frequently faced with patient discomfort and noncompliance. In the case of proteins, susceptibility to proteolysis and colloidal instability are additional difficulties. Consequently, a high drug concentration or a high dosing frequency becomes necessary, which may lead to adverse side effects [68-71]. Thus, drug delivery systems (DDS) are urgently needed, for the enhancement of the biopharmaceuticals bioavailability and selectivity, enabling a targeted controlled release profile.

Aiming at achieving an effective protein delivery, carriers such as liposomes, polymer micelles, and micro- or NPs have been developed [68, 70, 72-76]. Among them, nanometer-sized (<100 nm) polymer hydrogels have attracted growing interest. By trapping proteins in a hydrated polymer-network, these nanogels minimize denaturation, simultaneously allowing a slow, continuous and controlled release of the protein, maintaining an effective concentration for the necessary period of time [73, 77-79]. Nanocarriers enable localized and specific targeting to their intended tissues or cells, thereby decreasing drug doses and improving patient compliance [72]. Among nanosized systems, CS NPs have been tested as carriers for proteins, peptides and oligosaccharides.

Usually, proteins and vaccines are delivered via parenteral routes, due to their low bioavailability and/or poor immunogenicity when administered via non-parenteral routes [80]. In recent years, substantial progresses have been made on the use of non-invasive routes, such as mucosal (oral, nasal, pulmonary and colon) and transdermal, for the delivery of proteins and vaccines, yielding better patient compliance [80-83].

Oral Delivery

The oral route is considered the most convenient and comfortable means of drug administration, because of its non-invasive nature. It reduces the risk of infection, and do not require trained personnel. Drug molecules may cross the intestinal epithelium by transcellular or paracellular pathways. However, the bioavailability of orally administered proteins is usually poor, because of the hostile gastric and intestinal environments, and also the limited gastrointestinal (GI) mucosal permeability. The intestinal epithelium is a major barrier to the absorption of hydrophilic macromolecules, because they cannot diffuse across the lipid bilayer of the cell membranes, given the large Mw and hydrophilicity [84, 85]. In order to facilitate the paracellular transport of hydrophilic macromolecules, efforts have been made to induce transient opening of intercellular tight junctions [86, 87]. It has been reported that CS is able to enhance absorption in the intestinal lumen. A combination of mucoadhesion and a transient opening of the tight junctions in the mucosal membrane explain the enhanced absorption ability of CS. The mucoadhesive properties are attributed to an interaction between the positively charged CS and the negatively charged sialic-acid groups in mucins, which provides a prolonged contact time between the protein and the mucosal surface, thereby promoting its absorption [88, 89]. It has also been suggested that interactions of the CS positively charged amino groups, with the negatively charged cell surfaces and tight junctions, induce a redistribution of F-actin and tight junction's protein ZO-1, which triggers the increased paracellular permeability.

Lin et al. [90] reported the production, by ionic gelation, of NPs composed of CS and poly-γ-glutamic acid (γ-PGA) for insulin delivery. In the GI tract, the pH varies from acidic in the stomach to slightly alkaline in the small intestine. The fasting pH of the stomach is about 2.5 to 3.7, but in the presence of food, the pH drops to about 1.0 to 2.0, due to hydrochloric acid secreted by parietal cells. The proximal part of the small intestine (duodenum) has a pH value of about 6.0–6.6 due to neutralization of the acid by bicarbonates secreted by the duodenal mucosa and pancreas. The jejunum and ileum are the middle portion and terminal part of the small intestine, respectively, and their pH values are about 7.4. Therefore, characterization of NPs at distinct pH values, simulating the environments of the GI tract, was investigated. The stability and functionality of NPs *in vitro*, using Caco-2 cell monolayers, and *in vivo*, in a rat model were studied. The pKa values of CS and γ-PGA are 6.5 and 2.9, respectively. In the range of pH 2.5–6.6, CS and γ-PGA are ionized. The ionized **CS and γ**-PGA can form polyelectrolyte complexes, which results in a matrix structure with a spherical shape. Outside of this pH range, the NPs become unstable and subsequently broken apart. This is because, at **pH 1.2,** γ-PGA is not charged. Therefore, the little electrostatic interaction between the positively charged CS and the **neutral** γ-PGA causes the instability of NPs. Similarly, at pH 7.4, CS is neutral and thus NPs collapse. The authors observed that the CS NPs could transiently and reversibly open the tight junctions between Caco-2 cells, thus enhancing the paracellular permeability. However, the CS NPs at pH 7.4 appear to be less effective in opening tight junctions than at pH 6.6, due to the less positively charged CS. It was suggested that the electrostatic interaction between the positively charged CS and the negatively charged sites of ZO-1 proteins on cell surfaces induces a redistribution of cellular F-actin, leading to an increase in paracellular permeability.

The pH sensitivity and functionality of the CS NPs were confirmed in an animal study. At the duodenum (pH 6.0–6.6), while adhering and infiltrating into the mucus layer, the orally administered NPs may be degraded due to the presence of digestive enzymes in the intestinal fluids. Additionally, the pH environment may be changed (becoming neutral) while the NPs were infiltrating into the mucosa layer and approaching the intestinal epithelial cells. This may further lead to the collapse of NPs due to the change in the exposed pH environment. The dissociated CS from the degraded/collapsed NPs is then able to interact and modulate the function of ZO-1 proteins between epithelial cells. ZO-1 proteins are thought to link the occludin and actin cytoskeleton, playing important roles in the rearrangement of cell–cell contacts. Oral intake of NPs/insulin demonstrated a sustained decrease of the blood glucose level over a long period of time, at least 10 h. In a further development of this work, Sonaje et al. [91] prepared self-assembled NPs, by mixing γ-PGA with CS in the presence of $MgSO_4$ and TPP. The introduction of $MgSO_4$ in the preparation of CS NPs improved the stability in a broader pH range. The intestinal paracellular transport of insulin was investigated using Caco-2 cell monolayers. Additionally, the efficacy of NPs for oral delivery and intestinal absorption of insulin was investigated, in a diabetic rat model. The *in vitro* results showed that the mucoadhesive properties of CS NPs are affected by the pH and additionally, the transport of insulin across Caco-2 cell monolayers is pH-dependent: with increasing pH, the amount of insulin transported decreased significantly, due to the lower positive surface charge of the NPs, hence lower mucoadhesive and absorption enhancement ability. In addition, oral administration of insulin-loaded NPs demonstrated a significant hypoglycemic action for at least 10 hours, in diabetic rats.

Based on the work of Lin et al. [90] and Sonaje et al. [91], a mechanism for the paracellular delivery of insulin through the GI tract using CS NPs was proposed (Figure 5).

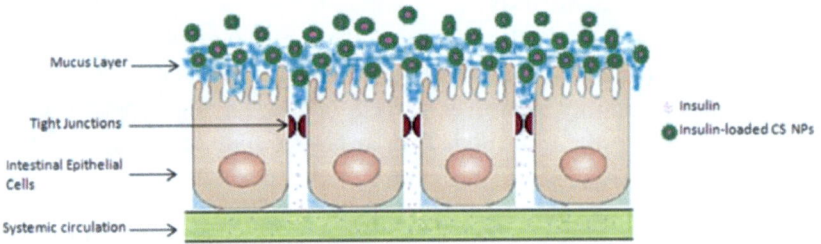

Figure 5. Schematic illustration of the hypothetic mechanism of the paracellular transport of insulin released from CS NPs via oral administration. NPs adhere and infiltrate into the mucus layer of intestinal epithelium. The infiltrated NPs disintegrate due the near neutral pH and release the loaded insulin; simultaneously, the CS opens the tight junctions allowing the insulin permeation through the paracellular pathway.

Despite some encouraging results, the poor solubility at physiological pH is a limitation for a more effective use of CS based NPs. Indeed, it has been shown that only protonated – soluble – CS can trigger the opening of the tight junctions, facilitating the paracellular transport of hydrophilic compounds [92]. The pH of the intestinal lumen is higher than the pKa of CS (6.5), limiting its efficiency as an absorption enhancer and suitability for protein delivery in neutral and physiological environments.

Several strategies have been attempted to overcome these drawbacks. Qian et al. [93] prepared hydrophilic CS NPs, by free radical polymerization of methyl methacrylate (MMA), N-dimethylaminoethyl methacrylate hydrochloride (DMAEMC), or N-trimethylaminoathyl methacrylate chloride (TMAEMC), which show a higher solubility in a broader pH range. These graft copolymer NPs enhanced the absorption and improved the bioavailability of insulin via the GI tract of normal male Sprague-Dawley rats. Trimethyl chitosan (TMC) NPs were also developed by other authors for insulin delivery [94, 95]. In an attempt to combine the mucoadhesion and the permeation enhancing effects of TMC and thiolated polymers, Yin et al. [96] synthetized a TMC-cysteine (TMC-Cys) conjugate. Thiolated polymers have been developed as a category of mucoadhesive polymers with reactive thiol groups immobilized on the polymeric structure. They can tightly and long lasting adhere to the intestinal mucus layer, through covalent bonding with mucin glycoproteins, via thiol-disulfide exchange reactions, hence providing a steep drug concentration gradient at the absorption sites and exerting an additional permeation enhancing effect. However, thiolated CS is also insoluble at physiological pH, which again restricts its application. TMC-Cys/insulin NPs were prepared through self-assembly driven by the electrostatic interaction between oppositely charged TMC-Cys and insulin. The authors [96] confirmed that, when reaching the small intestine, the positively charged TMC-Cys NPs are directed to the mucus layer, through electrostatic interaction with negatively charged sialic acid residues on mucin glycoproteins. Meanwhile, the free thiol groups on TMC-Cys are oxidized at neutral pH and disulfide bonds allowed to form between TMC-Cys and cysteine-rich mucin, contributing to the notably higher amount of TMC-Cys NPs immobilized in the mucus layer. So, mucoadhesion and permeation enhancing effects are significantly enhanced by the presence of cystein conjugates. Besides being biocompatible, TMC-Cys NPs showed a more potent hypoglycemic effect following both oral and ileal administration in normal rats than TMC NPs.

Nasal Delivery

The nasal mucosa is an attractive route for the delivery of vaccines because it has a relatively large absorptive surface and low proteolytic activity [82, 97-101]. Importantly, nasally administered vaccines can induce both local and systemic immune responses. However, most proteins are not well absorbed from the nasal cavity when administered as simple solutions. The major factors limiting the absorption of nasally administered proteins are the poor ability to cross the nasal epithelia, and the mucociliary clearance, which rapidly removes protein solutions from the absorption site [82, 83, 97]. Mucoadhesive, hydrophilic NPs have received much attention to overcome these obstacles and deliver protein antigens via the nasal route, because they strongly attach the mucosa increasing mucin viscosity [83, 97, 99, 100]. By this means, mucoadhesive NPs are able to decrease the nasal mucociliary clearance rate and thus increase the residence time of the formulation in the nasal cavity [82, 97].

Amidi and colleagues [102] prepared and characterized protein loaded TMC NPs as a nasal delivery system, by ionic crosslinking a TMC solution (with or without ovalbumin, the model protein studied) with TPP. It was observed that TMC NPs have a high loading efficiency (fraction of protein loaded) and capacity (amount of protein loaded per NPs dry weight) up to 50% (w/w). The integrity of the entrapped ovalbumin was preserved and release studies showed that more than 70% of the protein remained associated with the NPs for at least 3h of incubation in PBS (pH 7.4), at 37°C. Regarding biocompatibility, the NPs were non-cytotoxic, whereas a partially reversible cilio-inhibiting effect on the chicken trachea was observed. *In vivo* uptake studies indicated the transport of the protein across the nasal mucosa. Altogether, the authors concluded that TMC NPs are a promising nasal delivery system for proteins.

Other authors tested CS NPs as a nasal delivery system for insulin. Zhang *et al.* [103] used polyethylene glycol-grafted CS NPs to improve the systemic absorption of insulin, following nasal administration. The NPs were prepared by ionotropic gelation using TPP ions as the crosslinking agent. *In vitro* release studies showed an initial burst, followed by a slow release of insulin. Intranasal administration of the NPs in rabbits enhanced the absorption of insulin to a greater extent than the free protein. The nasal delivery of insulin using chitosan-acetyl-L-cysteine (CS-NAC) NPs was proposed by Wang *et al.* [104]. These authors observed that intranasal administration of CS-NAC NPs in rats enhanced the absorption of insulin by the nasal mucosa, as compared with unmodified CS NPs and control free insulin solution.

Mucosal vaccine strategies have emerged as a viable and attractive alternative to parenteral immunization. Advantages associated with mucosal vaccination are numerous: easy and low cost of administration, patient compliance, avoidance of the hepatic first pass metabolism and ability to induce mucosal as well as systemic immunity. Furthermore, the immune response generated at one mucosal site is able to induce a strong immune response at distal mucosal surfaces [105]. Westerink *et al.* [106] examined the effect of mucosal administration of tetanus toxoid (TT) in the presence of a non-ionic copolymer, Pluronic® F127 (F127) with CS or lysophosphatidylcholine (LPC), on the systemic and mucosal immune response. The results suggest that the two components of F127/CS appear to exert an additive or synergistic effect on the immune response.

Pulmonary Delivery

Pulmonary drug delivery for both local and systemic treatments has many advantages over other delivery routes. The lungs have a large surface area (43 to 102 m^2) [107]. In addition, mucociliary clearance is slower at the alveoli of the lungs than in the airways. Furthermore, the epithelia is thinner and more permeable, making possible the systemic absorption of peptides and proteins. Indeed, a number of high Mw drugs were demonstrated to be absorbed successfully through the lungs [107-109]. The successful delivery of the inhaled particles depends mostly on their size and density, and hence, on the aerodynamic diameter. The respirable fraction of these powders, generally the fraction of particles with an aerodynamic diameter ranging from 1 to 5 µm, should be as high as possible, as to guarantee a maximum deposition in the deep lung [110]. Independently of the method used to produce the aerosol, before reaching the deep lung, inhaled particles must overcome certain obstacles and lung defense mechanisms, essentially the effect of the branched airways structure and the mucus layer, which protects the epithelium in the tracheobronchial region. Particles targeted to the deep lung should be small enough to pass through the mouth, throat and conducting airways; however, if too small, they mail fail to deposit, being breathed out again. Therefore, they should have an aerodynamic diameter between 1 and 5 µm. Even so, a certain number of particles will be transported away from the lung by mucociliary clearance [111]. Once in the deep lung, particles will have to face at least two other defense mechanisms: the alveolar macrophages, and the enzymatic activity. The alveolar surface is covered by a thin layer of fluid with suspended macrophages, which play an important role in the lung defense. With the capacity of moving freely in the surface, they are able to engulf "foreign" substances from the airway surface, eliminating potentially damaging agents [111]. There is no consensus concerning the ideal size range to avoid or delay phagocytosis; however, it has been reported that the phagocytic activity is maximum for particles of 1–2 µm, decreasing for both smaller and larger particles out of this range [112]. Generally, authors agree that for particles in the micrometer range, the smaller the particle size, the higher is the probability of being captured [112]. Concerning the second defense mechanism (enzymatic activity), it is known that the lung presents a lower enzymatic activity when compared to other mucosal surfaces, such as the gastric. However, some enzymes have already been identified, as protease inhibitors, isozymes of the cytochrome P-450 family and lysozyme [113].

Grenha *et al.* [114] reported the preparation and characterization of dry powders containing protein-loaded TPP-CS-NPs, using aerosol excipients. Bovine insulin was chosen as model protein, with mannitol and lactose as excipients. The results showed that the obtained microspheres are mostly spherical and possess appropriate aerodynamic properties for pulmonary delivery. These NPs showed a good protein loading capacity, providing the *in vitro* release of 75–80% insulin within 15 min, and could be easily recovered from microspheres after contact with an aqueous medium, with no significant changes in their size and zeta potential values. Therefore, protein-loaded CS NPs could be successfully incorporated into microspheres with adequate characteristics as to reach the deep lung; in contact with the aqueous environment, the microspheres were able to release the NPs and then the therapeutic macromolecule.

Colon Delivery

Colon targeted drug delivery is useful in improving the absorption of peptide drugs via the GI tract. Site specific drug delivery to the colon is of special interest for drugs instable in the upper part of the GI tract, because of the peptidase activity in the small intestine. The colon is thought to have lower enzymatic activity than other regions of the GI, hence a greater absorption efficiency in this region would be expected, as long as the proteins/peptides are released locally [115]. Due to negligible activity of brush-border membrane and much less activity of peptidases and pancreatic enzymes, the colon has been considered suitable for the delivery of peptides and proteins.

Bayat *et al.* [116] developed a nanoparticulate system using two new quarternized derivatives of CS, triethylchitosan (TEC) and dimethylethylchitosan (DMEC), for insulin colon delivery. Insulin NPs (CS, TEC or DMEC NPs) were prepared by the polyelectrolyte complexation method. The three kinds of NPs showed a positive charge that could facilitate insulin uptake, allowing a low bursting effect and a steady release of insulin *in vitro*. DMEC NPs and TEC NPs had smaller particle size, higher insulin loading capacity and improved transport and absorption of insulin in GI tract, as compared with CS NPs. The blood glucose lowering effect of TEC NPs and DMEC NPs, after injection into ascending colon, was superior to that obtained with free insulin or CS NPs. This study indicated that NPs prepared from quaternized derivatives of CS might be a promising vehicle of administration of proteins and peptides via colon absorption.

Numerous studies highlight the importance of CS NPs for protein, peptide and oligosaccharide delivery, as summarized in table 2.

Table 2. Representative examples of the use CS NPs for proteins, peptides, and oligosaccharide delivery

Route of administration	Method of preparation	Remarks	Ref.
Therapeutic agent / Associated disease: Insulin / Diabetes Mellitus			
Oral delivery	NPs, composed of hydrophilic CS and γ-PGA.	Oral administration of insulin loaded NPs in diabetic rats demonstrated a sustained effect of decreasing the blood glucose level.	[90]
	Insulin-loaded CS NPs prepared by ionotropic gelation with TPP.	CS NPs enhanced the intestinal absorption of insulin.	[117]
Pulmonary delivery	Low-molecular-weight CS NPs prepared by solvent evaporation method.	*In vivo* administration of CS NPs containing insulin showed hypoglycemic activity.	[118]
Therapeutic agent/Associated disease: Calcitonin / Osteoporosis			
Oral delivery	CS-PEG nanocapsules obtained by the solvent displacement technique.	*In vivo* studies showed capacity of CS-PEG nanocapsules to enhance and prolong the intestinal absorption of salmon calcitonin.	[119]
Pulmonary delivery	Surface-modified DL-lactide/glycolide copolymer (PLGA) NPs with CS prepared by the emulsion solvent diffusion method.	CS-modified PLGA NPs loaded with elcatonin reduced blood calcium levels to 80% of the initial concentration and prolonged the pharmacological action to 24 h.	[120]
Therapeutic agent / Effect: Heparin / Anticoagulant and anti-asthmatic			
Oral delivery	NPs prepared by ionic gelation method without chemically modifying heparin.	No significant anticoagulant activity was detected after administration of the free heparin solution in a rat model, while administration of NPs was effective in the delivery of heparin into the blood stream	[121]
Pulmonary delivery	CS and mixtures of hyaluronic acid (HA) with heparin combined to form NPs by the ionotropic gelation technique.	Fluorescent heparin-loaded CS-HA NPs were effectively internalized by rat mast cells. *Ex vivo* experiments evaluated the capacity of heparin to prevent histamine release in rat mast cells indicating that the free or encapsulated drug exhibited the same dose–response behavior.	[122]

Therapeutic agent / Effect: Cyclosporin-A (CyA) / Immunosuppressant and extraocular disorders			
Ocular delivery	CyA-loaded CS NPs obtained using the ionic gelation technique.	*In vivo* experiments showed that, following topical instillation of CyA loaded CS NPs to rabbits, it was possible to achieve therapeutic concentrations in the external ocular tissues (i.e., cornea and conjunctiva) while maintaining negligible or undetectable CyA levels in the inner ocular structures.	[123]
Therapeutic agent/Associated disease: Prolidase / Prolidase deficiency (PD)			
Parenteral administration	CS NPs prepared by combining ionotropic gelation, with TPP, and ultrasonication treatment.	*Ex vivo* experiments performed by incubating different amounts of prolidase loaded CS NPs with skin human fibroblasts from PD patients for scheduled times.	[124]
	CS NPs loaded with PEGylated prolidase, obtained by combining ionotropic gelation and ultrasonication treatment.	The *ex vivo* evaluation of prolidase activity on PD fibroblasts indicated a good level of prolidase activity replacement up to 10 days.	[125]
Therapeutic agent / Effect: RGD / Anti-carcinogenic			
Intratumoral administration	Hydrophobically modified glycol CS (HGC) NPs containing a cyclic RGD peptide (RGD-HGC) prepared by a solvent evaporation method.	Intratumoral administration of RGD-HGC significantly decreased tumor growth and microvessel density compared to native RGD peptide injected either intravenously or intratumorally.	[126]
Therapeutic agent / Purpose: Tetanus toxoid / Tetanus vaccination			
Nasal administration	PEG-coated poly(lactic acid) (PLA) NPs, CS-coated poly(lactic acid–glycolic acid) (PLGA) NPs, and CS NPs prepared by ionotropic gelation with TPP.	The coating of PLGA NPs with the mucoadhesive polymer CS improved the stability of the particles in the presence of lysozyme. Moreover, these particles were very efficient in improving the local and systemic immune responses to tetanus toxoid.	[127]
Therapeutic agent / Associated disease: Anti-neuroexcitation peptide (ANEP) / Neuroexcitation associated diseases			
Brain-targeting delivery	ANEP-loaded TMC NPs prepared by ionic crosslinking of TMC with TPP.	The results showed that the targetability of ANEP to brain was significantly increased by TMC NPs. Absorptive-mediated transcytosis was believed to be the main pathway for the brain-targeting of ANEP-TMC/NPs.	[128]

3.2. Gene Delivery

Gene therapy is an emerging field in medical and pharmaceutical sciences, a very promising one due its potential for the treatment of a wide range of diseases, both inborn and acquired, by replacing defective genes, substituting missing ones, or silencing unwanted gene expression. However, naked therapeutic genes are rapidly degraded by nucleases, showing poor and non-specific cellular uptake and also low transfection efficiency [129]. Hence, the development of safe and efficient gene carriers is primordial for the success of gene therapy. Gene delivery systems include viral vectors, cationic liposomes and polycationic complexes. In spite of high transfection efficiency, the immune and oncogenic responses associated to viral vectors limit their therapeutic applications *in vivo*. To overcome these limitations, non-viral delivery systems (cationic liposomes and cationic polymers) have been proposed as a safe alternative. Besides being easily synthesized in large-scale, these nanoparticulate systems are targetable, have low immune response and unlimited DNA packaging capacity [130]. Non-viral systems, based on cationic polymers bearing amine groups in their backbone, are now extensively used as gene carriers; they form stable complexes with DNA, keeping it safe from nuclease degradation, and readily interact with cells membrane. Among non-viral vectors, CS and its derivatives are good delivery systems for DNA, antisense oligodeoxynucleotides and siRNA. CS gathers a number of desirable characteristics: cationic charge, biodegradability, biocompatibility, low toxicity, mucoadhesivness and functional groups with targeting ability. On the other hand, its transfection efficiency is relatively low, compromising potential clinical application. Several studies have been carried out to elucidate the influence of the CS-based formulation parameters on the gene expression [129]. In this section, we review the state of the art of DNA and siRNA CS-based delivery systems [131].

3.2.1. Chitosan Features Influencing Transfection

Many reports highlight the potential of CS as an efficient polymeric gene carrier. At acidic pH, below the pKa, the primary amines in the CS backbone are positively charged and interact with negatively charged DNA leading to the spontaneous formation of nanosized complexes. Under neutral or alkaline conditions, CS is only slightly charged, thus the CS and DNA binding is stabilized mainly by hydrogen bonding and hydrophobic interactions [132].

DNA can be carried entrapped into the NPs by hydrophobic [63] or ionic interactions [133]. Numerous factors influence the stability and transfection efficiency of CS-based systems, including Mw, DD, charge ratio of CS to DNA (N/P ratio), pH, serum and additives, which will be discussed bellow.

The CS Mw has a major influence on the size of the NPs, the CS/DNA complex stability [134], the unpacking of DNA after endocytosis and thus, overall, on the transfection efficiency [135]. Huang et al. [136] reported that CS with low Mw (<20kDa) only poorly retains the DNA upon dilution, consequentially being less capable of protecting it from degradation by DNase and serum components, resulting in low transfection efficiencies. On the other hand, Köpping-Höggard et al. [137] developed highly effective non-viral gene delivery systems using CS oligomers (1.2 to 10 kDa). The easier dissociation of the polyplexes was reflected in a greater gene expression, when compared to the more stable high-molecular-weight ultrapure CS-DNA polyplexes. Consequently, an intermediate stability must be achieved: the optimal CS Mw that allow extracellular DNA protection (favored by high Mw) versus efficient intracellular DNA release (favored by low Mw), in order to optimize the levels of transfection [129].

Higher DD results in increased positive charge, hence greater DNA binding capacity and cellular uptake [129]. According to Köpping-Höggard et al. [138], only CS with high charge density form stable complexes with pDNA. Similarly, Huang et al. [136] observed that CS with low Mw or DD (46 kDa and 61%, respectively) did not retain DNA efficiently and showed poor cellular uptake. Also, Kiang et al. [139] concluded that the DNA binding efficacy decrease for DD <70%, resulting in low luciferase expression, due to the particle destabilization caused by the bulky acetyl groups in the polymer chains. Lavertu et al. [140] reported as well high levels of luciferase expression, equivalent to those obtained with positive controls (Lipofectamine ^{TM} and FuGENE 6), using CS formulations with DD>80% and low Mw (10kDa), at pH 6.5.

The N/P ratio is given by the stoichiometry of CS's nitrogen and DNA's phosphate. The surface charge of the polyplexes depends on the N/P ratio, which influences the particles ability to interact with the negatively charged cell membrane [141]. Ishii et al. [142] reported that the transfection efficiency increases for charge ratios of 3 and 5, decreasing for further higher values. Another study, developed by Kim et al. using galactosylated CS/DNA [143], showed that complete shielding of DNA occurs at charge ratio of 5, with no significant improvement in the range 5–20. Galactosylated CS/DNA complexes with charge ratio above 5 (slightly positive zeta potential) were suitable for effective gene transfer. The most enhanced stability was obtained at charge ratio 10, due to the prevention of self-aggregation.

It has been suggested that strong interactions between CS and DNA results in highly stable particles, thereby preventing dissociation within the cell and leading to the absence of DNA translation. Attempting to reduce the CS-DNA interaction, Douglas *et al.* [144] associated an anionic biopolymer (alginate, 12–80 kDa) with low Mw CS (10 kDa, 90% DD). The presence of alginate in the complexes improved the yield of cell transfection. With the same aim, Duceppe *et al.* [135] used NPs made of ultra low Mw chitosan (ULMWCh, <10 kDa)/Hyaluronic acid (HA), as a novel potential carrier for gene delivery. Addition of HA to the NP formulation improved transfection rate from 0,7 to 25%. Peng and colleagues [145] demonstrated that mixtures of CS, DNA and poly(gamma-glutamic acid) (g-PGA) in aqueous media lead to the formation of "compounded NPs", containing domains of CS/DNA and CS/g-PGA. With this internal structure, the compounded NPs might disintegrate into a number of even smaller sub-particles, after cellular internalization, improving the dissociation capacity of CS and DNA. Consequently, an improved transfection was obtained.

The transfection efficiency of the CS complexes is also dependent on the pH of the culture medium. A pH slightly below 7 is optimal to achieve a good balance between DNA association and dissociation [129]. A recent study highlights the importance of the polyplex stability under different pH [146]. At pH 5.5, unmodified CS was most efficient in the DNA condensation. Conversely, when the pH was adjusted to 7.4, in the presence of high ionic strength, the condensation was strongly compromised, due to the reduction of both the degree of ionization and solubility. The condensation of DNA with TMC (50kDa) and TMC grafted with polyethylenoglycol (PEG) is less sensitive to pH and, at neutral pH, to the ionic strength. Likewise, Kadyala *et al.* [147] reported that the higher rate of transport of CS-based NP in Caco-2 cell layers occur at pH 5.5. Increasing the pH decreases the transport efficiency by 3 and 10-folds, for pH 6.4 and 7.4 respectively, owing to the NPs aggregation.

A mandatory requirement for the *in vivo* therapeutic application of gene delivery systems is the stability in serum. According to Ishii *et al.* [148] the presence of 10–20% serum enhances transfection, higher concentrations of serum yielding poorer results. Probably, the optimal serum concentration is determined by the overall effect on cells.

3.2.2. Biological Barriers in Cell Transfection

The body defense barriers at the humoral and cellular level have evolved to efficiently prevent intrusion of exogenous entities. Thus, the transfer of foreign genetic material to cells is a most challenging area, implying a safe and efficient method to deliver therapeutic genes to target cells. The gene carriers must meet a number of physical-chemical requirements; namely, the nanocarrier should have the following properties: stability in biological fluids, access to the target cells and cellular uptake, endosomal escape ability, appropriate intracellular trafficking and unpacking of polyplexes, and nuclear transport (Figure 6) [149].

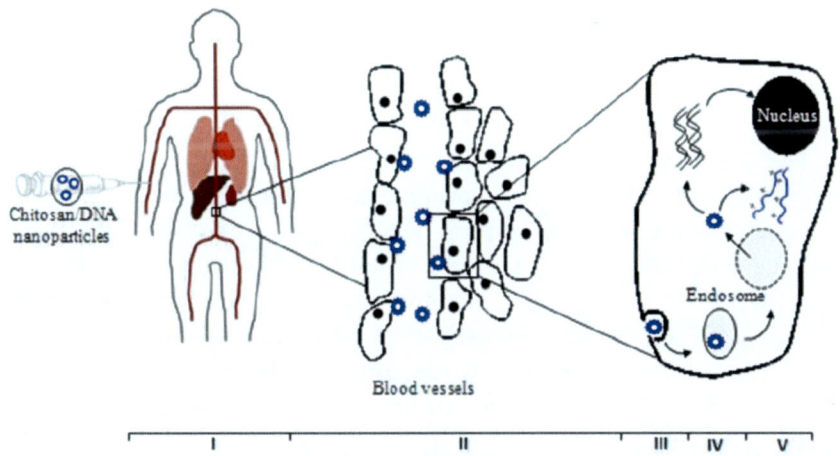

Figure 6. Biological barriers and gene delivery. Firstly, the polymeric complex should (I) be stable in the systemic circulation for a fairly long period of time, (II) able to access to the target cells and be internalized, (III) escape the endosomes to avoid degradation, (IV) reach the perinuclear space and allow unpacking of the DNA complexes and finally (V) translocation to the nucleus.

The cationic surfaces increase the solubility of the complexes in aqueous medium and facilitate its interaction with cells. However, when systemically administered, they readily interact with serum components, such as negatively charged serum albumin and other opsonins, originating the aggregation of particles in the blood stream through reduction of the zeta potential, thereby decreasing the ionic repulsion between particles. Opsonization results in the rapid clearance by cells of the mononuclear phagocytic system (MPS), specially macrophages in the liver (the Kupffer cells), spleen and bone marrow. In order to minimize the interactions with serum proteins, augmenting DNA survival in the bloodstream for the period of time necessary to reach the target tissue, hydrophilic polymers can be conjugated with CS (Figure 6, step I) [131]. PEGlyation of CS increases the systemic circulation time after intravenous administration, possibly by sterically avoiding the non-specific interactions between the serum-driven components and polyplexes [149]. Jiang *et al.* [150] synthesized a galactosylated poly(ethylene glycol)-chitosan-graft-polyethylenimine (Gal-PEG-CHI-g-PEI) as a potential hepatocyte-targeting gene carrier. After intravenous injection, PEI/DNA complexes rapidly accumulated in the lungs, whereas Gal-PEG-CHI-g-PEI/DNA complexes accumulated in the lungs, heart and liver, indicating that complexes had increased circulation time *in vivo* due to the hydrophilic group. Park and colleagues [151] introduced another hydrophilic group, poly(vinyl pyrrolidone) (PVP), into galactosylated CS to prevent aggregation of the complexes and the interaction with plasma proteins. The PVP surface effectively prevented the complex from interacting with albumin.

The polyplexes should reach the target cells without loss of integrity (Figure 6, step II). The positively charged particles readily attach to the cell surface via ionic interactions, thereby facilitating internalization by different endocytic mechanisms [152]. The cellular uptake of the polymeric polyplexes mostly occurs via non-specific adsorptive endocytosis. However, to improve cellular uptake efficiency and specificity, CS can be decorated with specific ligands (Table 3), which specifically recognize and bind receptors of the target cells (receptor-mediated endocytosis), improving the transfection efficiency [131].

Table 3. Representative examples of the CS and CS derivatives modified with cell specific ligands to improve the specificity to target cells

CS derivatives	Targeting ligands	Target cells	Remarks	Ref.
Folate-N-trimethyl CS (folate-TMC)	Folate	Cancer cells (KB and SKOV3) - folate receptor over-expressing) Lung cells (A549) and fibrosblast (NIH/3T3) - folate receptor deficient)	Folate conjugation increased the cellular uptake of the complex in KB cells and SKOV3 cells via folate receptor. The intracellular trafficking of the folate-TMC/pDNA was faster than TMC/pDNA due to the use of different trafficking pathways.	[153]
KNOB-CS	KNOB protein	Kidney cells (HEK293) Cancer cells (HeLa)	KNOB conjugated NPs improved gene expression level in HeLa and HEK293 cells by 130 and 7-folds, respectively.	[154]
Transferrin-High Mw (HMW) CS	Transferrin	Cancer cells (CaCo-2)	CS NPs decorated with transferrin enhanced the transport of NPs trough cell layers by 3- to 5-fold and led to higher stability of the NPs at higher pH.	[147]
Hyaluronic acid(HA)-CS	Hyaluronic acid	Corneal (HCE) and conjuctival (IOBA-NHC) cells	The endocytic process was mediated by hyaluronan receptor CD44	[155]
Mannosylated chitosan-graft-polyethylenimine (Man-CHI-g-PEI)	Mannose	Antigen presenting cells (APCs)	The transfection efficiency of Man-CHI-g-PEI/DNA complexes into macrophage cell line, which has mannose receptors, was higher than CHI-g-PEI as well as PEI.	[156]
Galactosylated poly(ethylene glycol)-chitosan-graft-polyethylenimine (Gal-PEG-CHI-g-PEI)	Galactose	Hepatocytes	Gal-PEG-CHI-g-PEI/DNA complexes transfected liver cells more effectively than PEI. Gal-CS is reported as hepatocyte-targeting gene carrier due to specific ligand–receptor interactions between galactose-moieties and asialoglycoprotein receptors (ASGPRs).	[150]

The introduction of the hydrophobic units in the CS-based complexes is also expected to increase transfection efficiency by modulating the complex interactions with cells, such as adsorption on cell surfaces and cellular uptake [131]. In addition, hydrophobic units in the polymeric carriers may assist in the dissociation of polymer/DNA complexes, facilitating the release of DNA which otherwise would be strongly bound through ionic interactions. Lee *et al.* [157] described the potential of thiolated CS for enhanced gene transfer. Indeed, the thiolated CS/pDNA nanocomplexes exhibited a gradual increase in mucin adsorption, probably due to the formation of covalent bonds between thiolated CS and cysteine-rich subdomains of mucin. *In vitro* and *in vivo* studies confirmed that thiolation of CS increases the transfection efficiency and sustained gene expression. The introduction of a trimethyltriazole group in C-6 position of CS allowed an increase in DNA binding ability, serum stability and significantly increase of cellular uptake, as compared to unmodified CS [158]. This effect was assigned to the ability of the trymethylammonium groups to open the tight junctions, leading to increased paracellular permeability and consequently higher transfection efficiency.

Once taken up by cells, via either adsorptive or receptor-mediated endocytosis, polyplexes are localized within the endosomal compartments, where pH rapidly drops to about 5 by the action of membrane bound ATP-driven proton pumps. The endosomes mature to lysosomes, where the arrested polyplexes disassemble due to the low pH and the released DNA may degrade. Therefore, the escape of polyplexes from the endosome is a critical step in the process (Figure 6, step III). Partially protonated polymers retain a substantial buffering capacity, which can lead to the protection of DNA from degradation. Since protons are diverted, the acidification of endosome is prevented; the continued action of the proton pump leads to the retention of chloride ions and therefore osmotic swelling occurs leading to subsequent endosome disruption [142]. Unfortunately, the buffering capacity of CS (pKa=6.5) is weak compared with PEI (pKa=8,7). Hence, CS have been frequently conjugated with PEI to take advantage of its proton sponge effect [159]. The combination of PEI with CS/DNA complex dramatically increased the luciferase expression in various cell lines, and the synergistic effect was proved to be induced by proton sponge effect of PEI [160]. However, CS-graft-PEI/DNA complexes frequently display higher transfection efficiency than PEI/DNA [161]. Köpping-Höggard *et al.* [138] studied in detail the effect of the addition of PEI to CS in the transfection efficiency. In this study, in contrast to PEI, ultrapure CS (UPC) displayed no buffering capacity at the acid endosomal pH-interval of 4.5–5.5, and thus the authors suggested the enzymatic degradation as a more likely mechanism for the endosomal escape of UPC polyplexes. Indeed, the enzymatic degradation products (oligo- and monosaccharides) may increase the osmolarity, followed by water influx, subsequent swelling and rupture of the vesicular membranes. It was stated that, whatever the mechanism, the efficiency of the PEI and UPC polyplexes depend on the rates at which the two polymers escape the endo/lysosomal compartment.

Several modifications of CS have been attempted to improve this effect. Imidazole-containing polymers have also been reported to act as a proton sponge, consequently enhancing the release of the complex into the cytoplasm following endocytosis. Kim et al. [162] used water soluble CS (WSC) conjugated with urocanic acid (UA) bearing an imidazole ring. The transfection efficiency was enhanced by grafting CS with UA, an effect that increases with the UA contents. Hu et al. [67] grafted hydrophobic moieties, stearic acid (SA), with CS oligosaccharide (CSO) (CSO–SA). Transfection using the CSO–SA/DNA complexes reached an efficiency of 15%, slightly below the figures obtained with Lipofectamine^{TM} (about 20%). The high transfection of CSO–SA/DNA complexes – as compared with CSO/DNA - is believed to rely on the chain of SA, which may favor the escape of CSO–SA/DNA complex from endosome. When the CSO–SA was used as a transfection carrier of pEGFP-C1, the fluorescence intensity increased gradually with the post-transfection time (until 76 h), and during this period cellular growth was observed. Conversely, a sharp increase on the transfection was detected with Lipofectamine^{TM}/DNA in 24h post-transfection, though after this period the DNA expression decreased rapidly, possibly due to the cytotoxicity of the formulation. The continuous, yet with relatively low efficiency, transfection of CSO–SA may be related to the slow rate of release of DNA. The pH sensitivity of the poly(propyl acrylic acid) can also be used to enhance the release of endocytosed drugs into the cytoplasmic compartment, because it exhibits maximal membrane disruption ability at pH 6. By incorporating this polymer in a CS gene carrier, Kiang et al. [163] confirmed the enhancement of the pDNA release from the endosomal compartment and improved gene expression.

After escaping from endosome, the complex should be able to unpack quickly, allowing the DNA to move towards the perinuclear space, where nuclear translocation of DNA takes place (Figure 6, step IV). While highly stable polyplexes may provide robust protection of DNA from extra- and intra-cellular nuclease attack, maximum transfection efficiency may not be achieved due to restriction in transcription. In contrast, polyplexes with lower stability may go through rapid uncoupling, causing premature degradation of plasmid

DNA in the cytosol. Therefore, maximum transfection efficiency may be achieved using a polymer with intermediate stability. The unpacking can be carried out within the endosome [164] or cytoplasm [165]. In the following step, the DNA or complex should move to the peri-nuclear space. To better understand the intracellular trafficking of pDNA/lactosylated-CS complexes, Hashimoto *et al.* [165] examined the effect of the endocytosis inhibitors on the transfection efficiency. Bafilomycin A1, a proton pump inhibitor, greatly depressed the luciferase activity of both pDNA/CS and pDNA/lactosylated-CS complexes. Monensin, an inhibitor of endosomal acidification, significantly decreased the gene expression of the pDNA/CS and pDNA/lactosylated-CS complexes. Nocodazole, which blocks transport from the early to late endosomes, resulted in the accumulation of cargo in the endosome compartment and improved transfection efficiency of the pDNA/CS complex, by about 2-3-fold. In contrast, in the case of pDNA/lactosylated-CS complexes, the transfection efficiency was decreased by nocodazole to 60% for HepG2 cells. Thus, the entrapment of pDNA/lactosylated-CS complexes in early endosome resulted in the obstruction of the release from the endosome. Although the transport of pDNA complexes to the late endosome/lysosome would raise the risk of the hydrolysis of DNA, the pDNA/lactosylated-CS complex showed high transfection efficiency, taking advantage of the release in perinuclear region.

Efficient nuclear localization of DNA is considered the final destination of gene delivery, since eukaryotic transcription is an essential intermediate step to convert genetic information into protein and is performed in the nucleus (Figure 6, step V). However, the mechanism of nuclear translocation of DNA from CS/DNA complexes is not yet understood [149].

3.2.3. DNA Therapy

Recent advances in gene delivery emphasize the application of CS-based NPs as gene carriers in cancer, rheumatoid arthritis, atherosclerosis, allergic asthma, tuberculosis, hemophilia A, hepatitis B, coxsackievirus B3 and respiratory syncytial virus (RSV) infections (Table 4), among others.

Table 4. Representative examples of the use CS/DNA NPs for gene therapy

Disease	Therapeutic agent	Administration	Remarks	Ref.
Colon adeno-carcinoma	IL-12 gene	Intratumoral	Mannosylated CS/pmIL-12 complexes administered in BALB/c mice bearing CT-26 tumor cells resulted in high expression levels of IL-12 and IFN-γ, suggesting that tumor growth was retarded due to the higher production of both cytokines. The IL-12 down-regulated angiogenesis and together with IFN-γ promotes apoptosis and cell cycle arrest.	[166]
Rheumatoid arthritis	IL-1 Ra gene	Intravenous	The human IL-1Ra remained in the serum of rats for 10 days and reverted the alterations in bone turnover (bone resorption versus formation) observed in arthritic animals.	[167]
Atherosclerosis	pCR-X8-HBc-CETP (pCETP)	Intranasal	Significant serum anti-CETP (cholesteryl ester transfer protein) IgG were detected and lasted for 21 weeks. The aortic lesions in the rabbits with NPs were lower than those treated with saline control. CS/pCETP NPs could significantly attenuate the progression of atherosclerosis.	[168]
Asthma	IFN-γ pDNA	Intranasal	CS/IFN-γ pDNA NPs (CIN) led to the normalization of airway inflammation and hyperresponsiveness (AHR), and return to normal lung morphology from the hyper-inflammatory condition induced by Ovalbumin (OVA) sensitization.	[169]
Tuberculosis	pDNA T-cell epitopes from *Mycobacterium tuberculosis*	Pulmonary	CS/DNA was able to induce the maturation of dendritic cells (DCs). pDNA incorporated in CS NPs induced increased levels of IFN-γ secretion compared to pDNA in solution.	[170]
Hemophilia A	FVIII pDNA	Oral	DNA polyplexes were detected in gastrointestinal tissues as well as in liver, spleen and additional systemic tissues in the hemophilia A mice. Functional Factor VIII protein was found in plasma reaching a level of 2-4% FVIII at day 22 after delivery.	[171]

Table 5. (Continued).

Disease	Therapeutic agent	Administration	Remarks	Ref.
Hepatitis B virus infection	pRc/CMV-HBs	Nasal	pRc/CMV-HBs loaded CS NPs resulted in serum anti-HBsAg and sIgA titers in the mucosal secretions. CS NPs were able to induce humoral mucosal and cellular immune responses.	[172]
Coxsackievirus B3 infection	pcDNA3-VP1 (encoding VP1, major structural protein of CVB3)	Intranasal	Mice immunized with CS/pcDNA3-VP1 produced higher levels of IgG and IgA. CS/DNA vaccine induced CVB3-specific systemic immunity (humoral and cellular) and protected mice from lethal CVB3 challenge.	[173]
Respiratory syncytial virus (RSV) infection	pDNA encoding a cytotoxic T-lymphocytes (CTL) epitope from M2 protein of RSV	Intranasal	Immunization with pDNA conjugated with CS induced *in vivo* peptide- and virus-specific CTL responses. In CS/DNA immunized mice a significant reduction in virus loaded in the lungs was observed.	[174]

3.2.4. siRNA Delivery

RNA-interference (RNAi) mediates knockdown of harmful or unwanted genes. In the RNAi process, double-stranded small interfering RNA (siRNA) with 21–23 nucleotides, endogenously produced or exogenously introduced, associates with a nucleic acid-protein complex called RNA-induced silencing complex (RISC). One of the strands is used to target a specific sequence in a particular messenger RNA (mRNA), leading to its degradation. Hence, the synthesis of the protein encoded by that mRNA molecule is prevented [175, 176]. The successful application of siRNA is largely dependent on the development of the delivery vehicle, due to its rapid degradation and poor cellular uptake *in vitro* and *in vivo* [177]. So, an ideal carrier for siRNA should be able to bind and condense siRNA, provide protection against degradation, specifically direct the siRNA to target cells, facilitate its intracellular uptake and escape from the endosome/lysosome into cytosol, and finally promote efficient gene silencing. Recently, Katas and Alper [178] were the first to explore the use of CS as polymeric carrier for siRNA delivery.

In comparison to usual DNA-based gene delivery, the extra vulnerability of RNA to enzymatic degradation represent additional hurdles to CS-mediated gene transfer, and make it even more challenging than conventional pDNA delivery. As the structure and size of siRNA are quite different from those of pDNA, the influence of the N/P ratio, serum and CS Mw, is also different; however, the effect of DD and pH is similar. Therefore, all parameters must be optimized specifically for the CS/siRNA complexes. NPs stability is required for extracellular siRNA protection, but disassembly is needed to allow RNA-mediated gene silencing through interaction with the intracellular RISC. An appropriate balance between protection and release of siRNA needs to be achieved, using a CS with the convenient Mw. Liu and colleagues [175] verified that CS with high Mw and DD result in the formation of discrete and stable NPs, 200 nm in size. CS/siRNA formulations prepared with low Mw CS (~10 kDa) showed almost no knockdown, whereas highest gene silencing efficiency (80%) was achieved using CS/siRNA NPs at N:P 150, with high Mw CS (114 and 170 kDa) and DD of 84% [137]. The influence of N/P ratio, here defined as the ratio of CS amino groups (N) to RNA phosphate groups (P), in the size of nanoparticle was described by Howard *et al.* [179], using a CS with Mw of 114 kDa. The nanoparticle hydrodynamic radius increased along with decreasing the N/P ratio. This suggests siRNA bridges the CS chains, higher concentrations leading to greater CS incorporation and possible interparticle aggregation. Liu *et al.* [175] investigated the influence of N/P ratio on the gene knockdown efficiency using chitosan/siRNA NPs (CS with 170 kDa and DD of 84%) in H1299 human lung carcinoma cells. It was found that the level of EGFP knockdown increased at higher N/P ratios (50 and 150), in comparison to low N/P (2 and 10) formulations; the NPs formed at N/P 150 showed the greatest level (80%) of EGFP knockdown. This result was explained by the increased nanoparticle stability at high N:P. Indeed, removal of excess polycations prior to transfection resulted in virtually no cellular knockdown, suggesting a possible role of a CS excess in the cellular permeation. Inside cells, the siRNA must be resistant to digestion by nucleases. Katas and Alper [178] studied the effect of serum in the free siRNA and CS–siRNA NPs stability and protection from nuclease degradation. After incubation with 5% FBS, free siRNA was intact only up to 30 min, being fully degraded after 48 h; on the other hand, siRNA in CS/TPP NPs started to degrade after 24 h incubation and full degradation was only observed after 72 h incubation. Interestingly, the siRNA recovered from CS–siRNA NPs was intact up to 7 h and fully degraded after 48 h incubation in 50% serum, while complete degradation of free siRNA was observed from the very first moments of incubation. Indeed, CS–siRNA NPs significantly protected siRNA from nuclease activity.

In order to improve the efficiency of RNA transfer using CS, several attempts on the vector improvement have been made over the past years. Katas and Alpar [178] synthesized CS NPs by ionic gelation using CS salts (CS hydrochloride and glutamate) and sodium TPP. Compared with standard chitosan, these NPs showed efficient siRNA transfer, which may be related to the higher RNA binding capacity and loading efficiency. Later, another group conjugated CS with thiamine pyrophosphate [180], a thiamine derivative. The CS-thiamine diphosphate-mediated siRNA silencing of endogenous EGFP gene occurred at best with 70–73% efficiency. This efficiency was associated with the increased nucleic acid binding ability and improved water solubility of the vector, due to the addition of extra amine groups from thiamine diphosphate and to the salt formation between the phosphate group of thiamine diphosphate and the amine group of CS. Lee *et al.* [181] synthesized CS NPs, by coacervation, to encapsulate siRNA in the presence of polyguluronate (PG), a block of guluronic acid residues present in the alginate backbone. The ability of PG to form stable NPs with CS was hypothesized, given its low Mw and ionic interactions with cations. CS/(PG+siRNA) NPs were the most efficient formulation to deliver siRNA into HEK 293FT and HeLa cells, as compared with NPs without PG or with alginate replacing PG. Ji and colleagues [182] developed CS/FHL2 siRNA NPs with a hydrodynamic radius of about 148 nm, which knock down about 67% FHL2 gene expression (over-expressed human colorectal cancer Lovo cells), very similar to the 69% reduced gene expression when siRNA was transfected with liposome Lipofectamine™. CS surface-modified poly(D,L-lactide-co-glycolide) (PLGA) nanospheres for siRNA delivery were prepared by Tahara *et al.* [183]. CS-PLGA nanospheres exhibited much higher encapsulation efficiency and were more effectively taken up by the cells than unmodified PLGA NPs, possibly due to electrostatics interactions with cell membrane. Consequently, the gene silencing efficiency of CS-PLGA nanospheres was higher and more prolonged.

3.2.5. Gene Silencing *In Vivo*

Only a few studies report the use of CS/siRNA for *in vivo* therapy. Some works described ahead in this review show the success of CS/siRNA NPs as a promising approach for the inhibition of gene expression *in vitro* and *in vivo*, and its therapeutic potential for the treatment of infections, allergic and inflammatory chronic diseases and cancer (Table 5).

Table 6 Examples of CS/siRNA complexes used in knockdown of gene expression

Disease	Therapeutic agent	Administration	Remarks	Ref.
Respiratory syncytial virus (RSV) infection	RSV-NS1 gene siRNA (siNS1)	Intranasal	Treatment of rats with siNS1 prior to RSV exposure reduced virus titers in the lung and prevented the inflammation and airway hyperresponsiveness (AHR) associated with the infection and development of asthma.	[184]
Asthma	Imiquimod and natriuretic peptide receptor A siRNA (siNPRA)	Transdermal	Imiquimod cream containing siNPRA CS NPs showed significantly reduced AHR, eosinophilia, lung histopathology and pro-inflammatory cytokines IL-4 and IL-5 in lung homogenates compared to controls.	[185]
Rheumatoid arthritis	anti-TNF-α Dicer-substrate siRNA (DsiRNA)	Intraperitoneal	CS NPs containing siRNA mediated TNF-α knockdown in primary peritoneal macrophages. Histological analysis of joints revealed minimal cartilage destruction and inflammatory cell infiltration in anti-TNF-α-treated mice.	[186]
Lung cancer	onco-protein Akt1 siRNA (siAkt)	Nasal	CS-graft-polyethylenimine carrier efficiently delivered siAkt and silenced onco-protein Akt1.	[187]

3.3. Delivery of Low Molecular Weight Drugs

Currently, the research on NPs based drug delivery systems focus on the selection of nanoparticulate carriers for suitable drug release profiles and also on its surface decoration, aiming at improving the targeting ability and *in vivo* biodistribution [188]. These are crucial goals, since drugs often fail to get favorable clinical outcomes, due to instability and reduced bioavailability. Furthermore, drugs need to be protected from degradation in the biological environment. The bioactivity is often limited by the inability to cross biological barriers and to reach the target site, specially when intracellular or intranuclear sites of action must be reached [124]. In addition, significant amounts of the administrated drug may distribute over the healthy tissues or organs, often leading to severe side effects. Among other drug delivery strategies, a great deal of attention has been directed to CS NPs as promising systems able to improve drug bioavailability, modify pharmacokinetics and/or protect the encapsulated drug [189]. In fact, CS NPs improve transmucosal permeability, enhance transport through the paracellular pathway, due to the good bio- and mucoadhesive properties, and induce structural reorganization of tight junction [190]. CS derivatives have been designed to improve the properties of native CS. Chemical modifications of CS originate amphiphilicity, an important characteristic for the formation of self-assembled NPs, potentially suited for drug delivery applications. The hydrophobic cores of the NPs may act as reservoirs for a variety of bioactive substances [191]. CS-based NPs as delivery systems of low molecular weight drugs have attracted attention for cancer and organ-specific therapy.

3.3.1. Cancer-Targeted Drug Delivery

Most anticancer agents do not specifically target cancer cells, but also normal tissues, leading to adverse effects following systemic administration. In addition, anticancer drugs are often poorly soluble in water; thus, organic solvents or detergents are necessary for clinical applications, resulting in undesirable side effects such as venous irritation and respiratory distress. Therefore, in an attempt to circumvent these limitations, research efforts have been concentrated on the development of new nanoparticulate drug delivery systems able to encapsulate a large quantity of drugs. Targeted drug delivery using long-circulating particulate drug carriers, such as polymeric NPs of controlled size, has high potential to improve the cancer therapy, providing a selective effect owing to the concentration of drugs at the tumor site, through enhanced permeability and retention (EPR) effect (Figure 7) and allowing lower distribution in healthy tissues [192] (Figure 7).

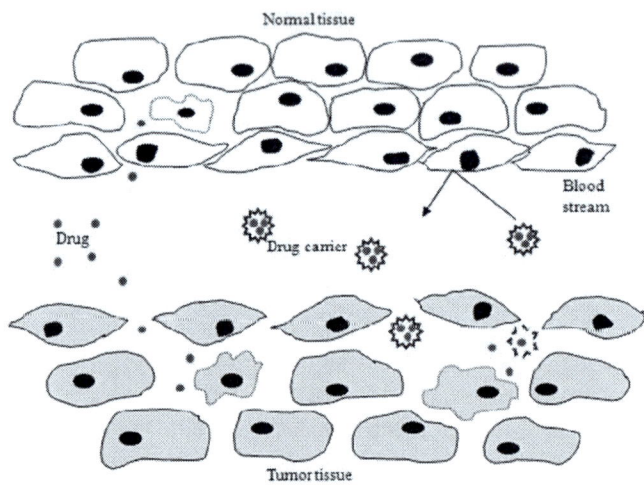

Figure 7. Schematic representation of anatomical differences between normal and tumor tissues. The defective tumor vasculature with disorganized endothelium allows passive targeting of nanoparticle carrier due EPR effect.

The use of polymeric NPs has been recognized as an effective strategy for passive tumor targeting, because its prolonged circulation time allows the accumulation and extravasation into the tumor tissue. The NPs diffuse through the anatomical and pathophysiological abnormalities of the tumor vasculature, utilizing the principle of EPR effect [193]. The size and surface of nanoparticle carriers play a crucial role in this process. Particles smaller than 200 nm and with hydrophilic surfaces tend to exhibit improved EPR effect, due to the increased residence time in the blood stream, which can only be possible avoiding opsonization.

The encapsulation of drugs into CS-based NPs can be physically achieved by hydrophobic interactions [194,195] or covalently binding to the polymer via a biodegradable spacer [196, 197] or using chemical cross-linkers [198, 199]. CS NPs have been investigated as carriers for diverse small molecular drugs, in recent years, as referred in Table 6.

Table 7. Examples of CS-based NPs systems for the delivery of low molecular weight drugs

Drug	CS derivative	Remarks	Ref.
Camptothecin (CPT)	Hydrophobically (5β-cholanic acid) modified glycol chitosan (HGC)	The CPT-HCG NPs intravenous injected exhibited significant antitumor effect and high tumor targeting ability towards human breast cancer xenografts, owing prolonged blood circulation time and high accumulation in tumors.	[200]
	poly(N-isopropylacrylamide) (NIPAAm)/CS	NIPAAm/CS NPs were sensitive to pH, which can be advantageous to target tumor cells. The CAMP loaded NPs drastically enhanced the cytotoxicity at pH 6.8	[201]
Doxorubicin (DOX)	CS-poly(acrylic acid) (PAA)	The DOX-loaded CS-PAA nanospheres remained in the systemic circulation for a longer period, compared with DOX solution. In mice liver, the DOX delivered from nanospheres was maintained constant at relatively high level	[202]
	Glycol CS	DOX loaded self-aggregates exhibited lower systemic toxicity but equivalent anti-tumor activity, compared to free DOX. The increased blood circulation time and sustained released resulted in the suppression of the tumor growth.	[203]
Paclitaxel (PTX)	Linoleic acid (LA) and poly(β-malic acid) (PMLA) grafted CS (LMCs)	The PTX-LMC1 was more effective in tumor suppression than free PTX, because the hydrophilic PLMA on the surface of LMC NPs decreased the uptake by the reticuloendothelial system.	[204]
	Hydrotropic oligomer-conjugated glycol CS (HO-GC)	HO-GC-PTX NPs presented a sustained drug release profile, rapid cellular uptake, and lower cytotoxicity. The PTX-encapsulated NPs were predominantly accumulated in the tumor tissue.	[205]
Cisplatin (CDDP)	HGC	CDDP-HGC NPs were successfully accumulated by tumor tissues in tumor-bearing mice, due to the prolonged circulation, enhanced permeability and EPR effect.	[206]

Table 7. (Continued).

Drug	CS derivative	Remarks	Ref.
Mitomycin C (MMC)	poly-ε-caprolactone coated with CS (CS-PCL)	CS-PCL NPs were selectively incorporated by bladder cancer cell line and MMC loaded NPs exhibited higher toxicity than free MMC.	[207]
5-fluorouracil (5-FU)	CS-polyaspartic acid (CTS-PASP)	The Bcl-2 gene family (regulatory factor group in apoptosis) was regulated by CTS-Pasp-5Fu and 5-Fu. The regulation effect of CTS-Pasp-5Fu NPs was more effective, enhancing the inhibition and inducing apoptosis of the gastric carcinoma.	[208]
Norcantharidin (NCTD)	Galactosylated CS (GC)	NCTD-GC NPs displayed tumor inhibition effect in mice	[209]
protophorphyrin IX (PpIX)	HGC	The released PpIX from NPs became highly phototoxic upon visible irradiation and in SCC7 tumor-bearing mice exhibited enhanced tumor specificity and increased therapeutic efficacy compared to free PpIX.	[210]
All-trans retinoic acid (ATRA)	Methoxy poly(ethylene glycol) (mPEG)-grafted CS (CP)	*In vitro* results suggested that tumor cell migration was most effectively inhibited by the polyion complex micelles (ATRA/CP) than ATRA free.	[211]
Docetaxel (DTX)	HGC	DTX-HGC NPs presented deformability, passing through the smaller pore size, owing their highly flexible features, prolonged circulation time, tumor targeting ability and higher antitumor efficacy.	[64]

The surface decoration of CS NPs with poly(ethylene glycol) (PEG) has attracted attention since it increases the physical stability of NPs, prolonging the circulation time in blood by avoiding the removal by the reticuloendothelial system and decreasing the positive charge of the particle surface. Hu *et al.* [212] verified that the PEGylation of stearic acid-grafted CS oligosaccharide (CSO-SA) micelles reduce significantly the cellular uptake by macrophages, not affecting the internalization by normal and tumor liver cells. A similar effect was observed by Qu *et al.* [213] using paclitaxel loaded mPEGOSC micelles (CS derivative with hydrophobic moieties of octyl and hydrophilic moieties of sulfate and polyethylene glycol monomethyl ether (mPEG) groups.

However, the accumulation of drugs in the tumor tissue is not always a guarantee of a successful therapy if the drug does misses the target site within the tumor cell, such as the cell membrane, cytosol, or nucleus. Park *et al.* [214] synthesized self-assembled NPs made of N-acetyl histidine-conjugated glycol CS (NAcHis-GC), a promising system for intracytoplasmic delivery of paclitaxel. Cellular uptake of NAcHis-GC NPs occurred by adsorptive endocytosis initiated by nonspecific interactions between NPs and cell membranes. Then, the NPs were exocytosed or localized in endosomes. In the slightly acidic environment of the endosomes, the drug-loaded NPs were disassembled due to breakdown of the hydrophilic/hydrophobic balance by the protonation of the imidazole group of NAcHis, providing a drug release into the cytosol. Therein, paclitaxel was effective in inducing arrest of cell growth. You *et al.* [65] developed another strategy for paclitaxel delivery. Micelles made of stearic acid and CS hold multiple hydrophobic "minor cores" near the surface, which improved the micelles internalization into cancer cells and accumulation of the drug in the cytoplasm. For antitumor drugs acting on the nucleus, effective internalization and nucleus accumulation is mandatory. Although, nuclear import of many nuclear proteins is based on the presence of a peptidic nuclear localization signals (NLS), other non-peptidic NLS exist, namely sugar molecules. You *et al.* [66] observed that the CS-g-stearic acid micelles loaded DOX presented an enhanced nuclear location comparing to free DOX, possibly due to the import of the micelles loaded drug occurring via a sugar-dependent manner.

Cancer cells often over-express specific antigens or receptors on the cell surface that can be used for active targeting. Chemical modification of the drug carrier using targeting moieties can precisely direct NPs to receptors on the tumor tissue [191]. For successful active targeting, the specific receptors should be expressed exclusively on the cancer cells. The targeting moieties most used are galactose, transferrin (Tf), folic and hyaluronic acids (HA). Glycotargeting takes advantage of a highly specific interaction between the carbohydrate ligands conjugated on macromolecules and the endogenous lectins present on the targeted cells. Because of their high density on the surface of hepatoma cells in the liver cancers, the asialoglycoprotein (ASGP) receptors are a particularly attractive site for glycotargeting. Among the glycoconjugated macromolecules, galactosylated CS was found to be a suitable material for liver-targeting drug/gene delivery [215]. Mi et al. [215] confirmed that the galactosylated CS NPs had higher specific interaction with hepatoma cells than CS NPs, via the ligand-receptor (ASPG)-mediated recognition, leading to a high affinity to HepG2 cells. The Tf is also over-expressed in tumor tissues, hence it can be used as a ligand for tumor targeting. Tf was covalently bound to the Dox-loaded palmitoylated glycol CS (GCP) vesicles by Dufes et al. [216]. The Tf decorated vesicles were taken up faster (after 1–2 h) and DOX reached the nucleus after 60–90 min, leading to higher cytotoxicity than GCP DOX *in vitro*, although this good *in vitro* performance did not translate into a therapeutic advantage *in vivo*. All vesicles reduced the tumor size on day 2, but were, overall, less active than the free drug. Folate receptors (FRs) are also frequently over-expressed in human epithelial cancerous cells. Therefore, folate-conjugated drugs or carriers can be rapidly internalized into cancer cells via receptor-mediated endocytosis. Folate-conjugated stearic acid grafted CS (54 kDa) NPs, produced by You et al. [217], were rapidly took up by Hela cells (over-expressed FRs) as compared to A549 cells (deficient FRs cell line). PTX was encapsulated into these micelles; the lethal half dose of taxol (a clinical formulation containing PTX) on A549 and Hela cells is 7.0 and 11.0 $\mu g\ ml^{-1}$, respectively while for PTX-loaded micelles these values were reduced to 0.32 $\mu g\ ml^{-1}$ and 0.268 $\mu g\ ml^{-1}$. These results were attributed to the increased intracellular delivery of the drug. Most malignant solid tumors and their surrounding stromal tissue contain elevated levels of HA, which can provide a matrix that facilitates invasion [218]. HA receptors, such CD44, are also over-expressed in tumor cells; indeed, cells with metastatic potential often show enhanced binding and internalization of HA. Jain and Jain [218] explored the utilization of HA grafted CS NPs for the effective delivery of 5-FU to colon tumors. HA-CS NPs showed significantly higher uptake by cancer cells, about 7.9 fold as compared to uncoupled NPs, which clearly indicate that the uptake of HA coupled NPs occurred via CD44 receptors of HT-29 cancer cells.

3.3.2. Organ-Specific Drug Delivery

NPs constitute a versatile drug delivery system, with the ability to overcome physiological barriers, guiding drugs to specific cells or intracellular compartments, either by passive or receptor-mediated targeting mechanisms. The NPs can be targeted to organs such as the eye, liver, spleen, lung and lymph; because of their very small size, they can pass through the narrowest capillaries [219].

Ocular Delivery

Efforts in ocular drug delivery have been made to improve the bioavailability and to prolong the residence time of drugs applied topically onto the eye [220]. Campus *et al.* [221] concluded that CS-coated poly-ε-caprolactone nanocapsules enhanced the penetration of an encapsulated dye through the cornea, probably due to the extended adhesion of the nanocapsules at the superficial layers of the epithelium. *In vivo* studies showed that the amounts of fluorescent CS in cornea and conjunctiva were significantly higher for fluorescent CS NPs than for a control fluorescent CS solution, these amounts being fairly constant for up to 24 h [222]. More recently, Motwani *et al.* [223] used mucoadhesive CS-sodium alginate (ALG) NPs as a new vehicle for the prolonged topical ophthalmic delivery of antibiotic gatifloxacin. Unfortunately, no evidence of the *in vitro* or *in vivo* behavior of this formulation has been reported. Badawi *et al.* [224] observed that, following topical instillation of a CS nanocarrier loaded with indomethacin (IM) to rabbits, it was possible to achieve therapeutic concentration of IM in the cornea and fairly high IM level in inner ocular structure. These levels were significantly higher than those obtained following instillation of IM solution. IM concentration delivered from nanocarriers in the cornea was sufficiently high to adequately suppress inflammatory process.

Liver-Target Drug Delivery

The diammonium glycyrrhizinate (DG) is used for the treatment of chronic hepatitis. Yang *et al.* [225] produced lactose-conjugated PEG-grafted-CS (Lac-PEG-g-CS) to promote liver-targeted delivery of DG, because lactose can be recognized by asialoglycoprotein receptor (ASGP-R) on the cell surface of liver. Indeed, the Lac-PEG-g-CS delivered DG more effectively to the liver than the PEG-g-CS micelles. Although reported as a therapeutic agent, the glycyrrhizin was also used for the surface decoration of the CS NPs (CS-NPs-GL), a novel hepatocyte-targeted delivery system [226]. CS-NPs-GL NPs were preferentially accumulated into hepatocytes and the cellular uptake was dependent on incubation time and dose of NPs, which indicated that the internalization of the NPs into hepatocytes was mostly mediated by a ligand-receptor interaction. Lin *et al.* [227] also described the conjugation of glycyrrhizin with N-caproyl CS.

Brain-Target Drug Delivery

The Alzheimer disease (AD) is a chronic neurodegenerative disorder accompanied by the gradual and progressive loss of functional and psychomotor abilities. The female sex hormone, 17β-estradiol, is involved in the regulation of brain development. Long-term oestrogen replacement has proved to be beneficial in the prevention and treatment of Alzheimer's disease [228]. Considering the therapeutic potential of E2, Wang *et al.* [228] studied the levels of estradiol (E2) in blood and the cerebrospinal fluid in rats following intranasal administration of E2-loaded CS NPs. The E2 was directly transported from the nasal cavity into the cerebrospinal fluid, in rats; the CS NPs are thus able to significantly improve the E2 transport to central nervous system. Later, the same group synthesized TMC surface-modified poly(D,L-lactide-co-glycolide) (PLGA) NPs (TMC/PLGA–NP) [229]. As a cationic ligand, TMC can facilitate the active transport of NPs via absorptive-mediated transcytosis (AMT) across the cerebral endothelium and, so TMC-modified NPs could be used as a drug carrier for brain delivery, overcoming the blood–brain barrier (BBB). PLGA–NP and TMC/PLGA–NP loaded 6-coumarin, as a probe, were injected into the caudal vein of mice; higher accumulation of TMC/PLGA–NP in the cortex and third ventricle was observed, as compared to PLGA–NP, demonstrating that the TMC/PLGA NPs pass through the endothelial cells of the BBB, reaching the brain parenchyma. This effect was confirmed by the behavior tests in AD transgenic mice, in which the neuroprotective effects of TMC/PLGA–NP loaded with coenzyme Q10 (Co Q10) were superior to the PLGA–NP and solution, and markedly improved the spatial memory.

Lung-Target Drug Delivery

In general, NPs delivery to the lungs is an attractive concept because retention of the particles in the lungs, accompanied with a prolonged drug release, can be achieved using large porous nanoparticle matrices. On the other hand, it has been shown that NPs uptake by alveolar macrophages can be reduced using particles smaller than 260 nm. Both effects combined might improve local pulmonary drug therapy [230]. Asthma is a chronic inflammatory disease of the airway characterized by the infiltration of eosinophils, epithelial hyperplasia leading to hypersecretion of mucus and the presence of airway hyperresponsiveness to a variety of stimuli. Theophylline was used for the treatment of asthma but side effects limit its application. Thus, Lee *et al.* [231] hypothesized that the absorption of theophylline through bronchial mucosa could be enhanced by administration with thiolated CS NPs (TCNs), because of their greater mucoadhesiveness and permeability properties. In an allergic asthma mouse model, intranasal delivery of theophylline complexed with TCNs augmented the anti-inflammatory effects of the drug compared to theophylline administered alone, or loaded into unmodified CS NPs.

Chapter IV

Conclusions

The application of CS nanoparticulate systems in drug delivery has great potential. Exciting concepts and sophisticated formulations have been produced using CS and its derivatives. However, several issues must be addressed such that these possibilities can be fully exploited and reach clinical application. A wider choice of pure, medical grade chitosan and its derivatives is needed. A better control over the stability of the NPs is necessary, in particular at physiological pH. In many cases, *in vitro* results are not reproduced *in vivo*, hence more knowledge on the fate of the CS NPs *in vivo* is mandatory. The interaction of NPs with serum proteins (in the blood), the biodistibution and intracellular trafficking must be more comprehensively characterized, as well as toxicological issues.

Acknowledgments

Paula Pereira, Vera Carvalho and Reinaldo Ramos were supported respectively by the grant SFRH/BD/64977/2009, SFRH/BD/27359/2006, SFRH/BD/27404 / 2006, from Fundação para a Ciência e Tecnologia (FCT), Portugal. This review was also supported by FCT through the project PTDC/BIO/67160/2006.

References

[1] Kas, H.S., Chitosan: properties, preparations and application to microparticulate systems. *J. Microencapsul.* 1997. 14(6): p. 689-711.

[2] Muzzarelli, R.A.A., Chitin / by Riccardo A. A. Muzzarelli. 1977, Oxford; New York : Pergamon Press.

[3] Chandy, T. and C.P. Sharma, Chitosan as a biomaterial. *Biomater. Artif. Cells Artif. Organs.* 1990. 18(1): p. 1-24.

[4] Sinha, V.R., et al., Chitosan microspheres as a potential carrier for drugs. *Int. J. Pharm.* 2004. 274(1-2): p. 1-33.

[5] Cho, Y.W., et al., Preparation and solubility in acid and water of partially deacetylated chitins. *Biomacromolecules.* 2000. 1(4): p. 609-614.

[6] He, P., S.S. Davis, and L. Illum, In vitro evaluation of the mucoadhesive properties of chitosan microspheres. *International Journal of Pharmaceutics.* 1998. 166(1): p. 75-88.

[7] Lehr, C.M., et al., Invitro Evaluation of Mucoadhesive Properties of Chitosan and Some Other Natural Polymers. *International Journal of Pharmaceutics.* 1992. 78(1): p. 43-48.

[8] Denkbas, E.B. and R.M. Ottenbrite, Perspectives on: Chitosan drug delivery systems based on their geometries. *Journal of Bioactive and Compatible Polymers.* 2006. 21(4): p. 351-368.

[9] Muzzarelli, R.A.A. and C. Muzzarelli, Chitosan chemistry: Relevance to the biomedical sciences. *Polysaccharides 1: Structure, Characterization and Use.* 2005. 186: p. 151-209.

[10] Biagini, G., et al., Morphological-Study of the Capsular Organization around Tissue Expanders Coated with N-Carboxybutyl Chitosan. *Biomaterials.* 1991. 12(3): p. 287-291.

[11] Chen, X.G., et al., The effect of carboxymethyl-chitosan on proliferation and collagen secretion of normal and keloid skin fibroblasts. *Biomaterials.* 2002. 23(23): p. 4609-4614.
[12] Mi, F.L., et al., Control of wound infections using a bilayer chitosan wound dressing with sustainable antibiotic delivery. *Journal of Biomedical Materials Research.* 2002. 59(3): p. 438-449.
[13] Mi, F.L., et al., Fabrication and characterization of a sponge-like asymmetric chitosan membrane as a wound dressing. *Biomaterials.* 2001. 22(2): p. 165-173.
[14] Ravi Kumar, M.N.V., A review of chitin and chitosan applications. *Reactive and Functional Polymers.* 2000. 46(1): p. 1-27.
[15] Hirano, S., Chitin and chitosan as novel biotechnological materials. *Polymer International.* 1999. 48(8): p. 732-734.
[16] Yokoyama, A., et al., Development of calcium phosphate cement using chitosan and citric acid for bone substitute materials. *Biomaterials.* 2002. 23(4): p. 1091-1101.
[17] Yamaguchi, I., et al., Preparation and mechanical properties of chitosan/hydroxyapatite nanocomposites. *Bioceramics.* 2000. 192-1: p. 673-676.
[18] Yamaguchi, I., et al., Preparation and microstructure analysis of chitosan/hydroxyapatite nanocomposites. *Journal of Biomedical Materials Research.* 2001. 55(1): p. 20-27.
[19] Zhang, Y. and M.Q. Zhang, Three-dimensional macroporous calcium phosphate bioceramics with nested chitosan sponges for load-bearing bone implants. *Journal of Biomedical Materials Research.* 2002. 61(1): p. 1-8.
[20] Haipeng, G., et al., Studies on nerve cell affinity of chitosan-derived materials. *J. Biomed. Mater Res.* 2000. 52(2): p. 285-295.
[21] Zhao, F., et al., Preparation and histological evaluation of biomimetic three-dimensional hydroxyapatite/chitosan-gelatin network composite scaffolds. *Biomaterials.* 2002. 23(15): p. 3227-3234.
[22] Zhang, Y. and M.Q. Zhang, Microstructural and mechanical characterization of chitosan scaffolds reinforced by calcium phosphates. *Journal of Non-Crystalline Solids.* 2001. 282(2-3): p. 159-164.
[23] Zhang, Y. and M.Q. Zhang, Synthesis and characterization of macroporous chitosan/calcium phosphate composite scaffolds for tissue engineering. *Journal of Biomedical Materials Research.* 2001. 55(3): p. 304-312.

[24] Muzzarelli, R., et al., Antimicrobial properties of N-carboxybutyl chitosan. *Antimicrob. Agents Chemother.* 1990. 34(10): p. 2019-2023.
[25] Hu, Y., et al., Self-aggregation of water-soluble chitosan and solubilization of thymol as an antimicrobial agent. *J. Biomed. Mater. Res.* A, 2009. 90(3): p. 874-881.
[26] Wilson, B., et al., Chitosan nanoparticles as a new delivery system for the anti-Alzheimer drug tacrine. *Nanomedicine.* 2010. 6(1): p.144-152.
[27] Sun, Y., et al., Magnetic chitosan nanoparticles as a drug delivery system for targeting photodynamic therapy. *Nanotechnology.* 2009. 20(13): p. 135102. doi: 10.1088/0957-4484/20/13/135102.
[28] Ta, H.T., et al., Chitosan-dibasic orthophosphate hydrogel: a potential drug delivery system. *Int. J. Pharm.* 2009. 371(1-2): p. 134-141.
[29] Wang, T., et al., Quaternized chitosan (QCS)/poly (aspartic acid) nanoparticles as a protein drug-delivery system. *Carbohydr. Res.* 2009. 344(7): p. 908-914.
[30] Kim, S., et al., Development of chitosan-ellagic acid films as a local drug delivery system to induce apoptotic death of human melanoma cells. *J. Biomed. Mater Res. B Appl. Biomater.* 2009. 90(1): p. 145-155.
[31] Noel, S.P., et al., Chitosan films: a potential local drug delivery system for antibiotics. *Clin. Orthop. Relat. Res.* 2008. 466(6): p. 1377-1382.
[32] Ciofani, G., et al., Alginate and chitosan particles as drug delivery system for cell therapy. *Biomed. Microdevices.* 2008. 10(2): p. 131-140.
[33] Krauland, A.H. and M.J. Alonso, Chitosan/cyclodextrin nanoparticles as macromolecular drug delivery system. *Int. J. Pharm.* 2007. 340(1-2): p. 134-142.
[34] Enriquez de Salamanca, A., et al., Chitosan nanoparticles as a potential drug delivery system for the ocular surface: toxicity, uptake mechanism and in vivo tolerance. *Invest. Ophthalmol. Vis. Sci.* 2006. 47(4): p. 1416-1425.
[35] Shimono, N., et al., Multiparticulate chitosan-dispersed system for drug delivery. *Chem. Pharm. Bull.* (Tokyo), 2003. 51(6): p. 620-624.
[36] Shimono, N., et al., Chitosan dispersed system for colon-specific drug delivery. *Int. J. Pharm.* 2002. 245(1-2): p. 45-54.
[37] Chen, A., C. Hou, and J. Bao, [Clinical study of gentamycin-loaded chitosan drug delivery system]. *Zhongguo Xiu Fu Chong Jian Wai Ke Za Zhi,* 1998. 12(6): p. 355-358.
[38] Bernkop-Schnurch, A., A. Krauland, and C. Valenta, Development and in vitro evaluation of a drug delivery system based on chitosan-EDTA BBI conjugate. *J. Drug Target.* 1998. 6(3): p. 207-214.

[39] Wilson, B., et al., Chitosan nanoparticles as a new delivery system for the anti-Alzheimer drug tacrine. *Nanomedicine-Nanotechnology Biology and Medicine.* 2010. 6(1): p. 144-152.

[40] Panyam, J. and V. Labhasetwar, Biodegradable nanoparticles for drug and gene delivery to cells and tissue. *Advanced Drug Delivery Reviews.* 2003. 55(3): p. 329-347.

[41] Nagarwal, R.C., et al., Polymeric nanoparticulate system: A potential approach for ocular drug delivery. *Journal of Controlled Release.* 2009. 136(1): p. 2-13.

[42] Thanoo, B.C., M.C. Sunny, and A. Jayakrishnan, Cross-Linked Chitosan Microspheres - Preparation and Evaluation as a Matrix for the Controlled Release of Pharmaceuticals. *Journal of Pharmacy and Pharmacology.* 1992. 44(4): p. 283-286.

[43] Akbuga, J. and G. Durmaz, Preparation and Evaluation of Cross-Linked Chitosan Microspheres Containing Furosemide. *International Journal of Pharmaceutics.* 1994. 111(3): p. 217-222.

[44] Jameela, S.R. and A. Jayakrishnan, Glutaraldehyde Cross-Linked Chitosan Microspheres as a Long-Acting Biodegradable Drug-Delivery Vehicle - Studies on the in-Vitro Release of Mitoxantrone and in-Vivo Degradation of Microspheres in Rat Muscle. *Biomaterials.* 1995. 16(10): p. 769-775.

[45] Pavanetto, F., et al., Evaluation of process parameters involved in chitosan microsphere preparation by the o/w/o multiple emulsion method. *Journal of Microencapsulation.* 1996. 13(6): p. 679-688.

[46] Hennink, W.E. and C.F. van Nostrum, Novel crosslinking methods to design hydrogels. *Advanced Drug Delivery Reviews.* 2002. 54(1): p. 13-36.

[47] Nishimura, K., et al., Macrophage activation with multi-porous beads prepared from partially deacetylated chitin. *J. Biomed. Mater Res.* 1986. 20(9): p. 1359-1372.

[48] Berthold, A., K. Cremer, and J. Kreuter, Preparation and characterization of chitosan microspheres as drug carrier for prednisolone sodium phosphate as model for anti-inflammatory drugs. *Journal of Controlled Release.* 1996. 39: p. 17-25.

[49] Tokumitsu, H., H. Ichikawa, and Y. Fukumori, Chitosan-gadopentetic acid complex nanoparticles for gadolinium neutron-capture therapy of cancer: preparation by novel emulsion-droplet coalescence technique and characterization. *Pharm. Res.* 1999. 16(12): p. 1830-1835.

[50] Fernandez-Urrusuno, R., et al., Enhancement of nasal absorption of insulin using chitosan nanoparticles. *Pharm. Res.* 1999. 16(10): p. 1576-1581.

[51] Calvo, P., et al., Chitosan and chitosan/ethylene oxide-propylene oxide block copolymer nanoparticles as novel carriers for proteins and vaccines. *Pharm. Res.* 1997. 14(10): p. 1431-1436.

[52] Bodmeier, R., H.G. Chen, and O. Paeratakul, A novel approach to the oral delivery of micro- or nanoparticles. *Pharm. Res.* 1989. 6(5): p. 413-417.

[53] Liu, C., et al., Preparations, characterizations and applications of chitosan-based nanoparticles. *Journal of Ocean University of China.* (English Edition), 2007. 6(3): p. 237-243.

[54] Mansouri, S., et al., Characterization of folate-chitosan-DNA nanoparticles for gene therapy. *Biomaterials.* 2006. 27(9): p. 2060-2065.

[55] Chen, F., Z.R. Zhang, and Y. Huang, Evaluation and modification of N-trimethyl chitosan chloride nanoparticles as protein carriers. *Int. J. Pharm.* 2007. 336(1): p. 166-173.

[56] Gan, Q. and T. Wang, Chitosan nanoparticle as protein delivery carrier--systematic examination of fabrication conditions for efficient loading and release. *Colloids Surf. B Biointerfaces.* 2007. 59(1): p. 24-34.

[57] Xu, Y. and Y. Du, Effect of molecular structure of chitosan on protein delivery properties of chitosan nanoparticles. *Int. J. Pharm.* 2003. 250(1): p. 215-226.

[58] Maitra, A., Determination of size parameters of water-Aerosol OT-oil reverse micelles from their nuclear magnetic resonance data. *The Journal of Physical Chemistry.* 1984. 88(21): p. 5122-5125.

[59] Agnihotri, S.A., N.N. Mallikarjuna, and T.M. Aminabhavi, Recent advances on chitosan-based micro- and nanoparticles in drug delivery. *J. Control Release.* 2004. 100(1): p. 5-28.

[60] de Moura, M.R., F.A. Aouada, and L.H.C. Mattoso, Preparation of chitosan nanoparticles using methacrylic acid. *Journal of Colloid and Interface Science.* 2008. 321(2): p. 477-483.

[61] Hu, Y., et al., Synthesis and characterization of chitosan-poly(acrylic acid) nanoparticles. *Biomaterials.* 2002. 23(15): p. 3193-3201.

[62] Gonçalves, C., P. Pereira, and M. Gama, Self-Assembled Hydrogel Nanoparticles for Drug Delivery Applications. *Materials.* 2010. 3(2): p. 1420-1460.

[63] Yoo, H.S., et al., Self-assembled nanoparticles containing hydrophobically modified glycol chitosan for gene delivery. *J. Control Release.* 2005. 103(1): p. 235-243.
[64] Hwang, H.Y., et al., Tumor targetability and antitumor effect of docetaxel-loaded hydrophobically modified glycol chitosan nanoparticles. *J. Control Release.* 2008. 128(1): p. 23-31.
[65] You, J., et al., Polymeric micelles with glycolipid-like structure and multiple hydrophobic domains for mediating molecular target delivery of paclitaxel. *Biomacromolecules.* 2007. 8(8): p. 2450-2456.
[66] You, J., et al., Improved cytotoxicity of doxorubicin by enhancing its nuclear delivery mediated via nanosized micelles. *Nanotechnology.* 2008. 19: p. 255103. doi: 10.1088/0957-4484/19/25/255103.
[67] Hu, F.Q., et al., A novel chitosan oligosaccharide-stearic acid micelles for gene delivery: properties and in vitro transfection studies. *Int. J. Pharm.* 2006. 315(1-2): p. 158-166.
[68] Haag, R. and F. Kratz, Polymer therapeutics: concepts and applications. *Angew. Chem. Int. Ed. Engl.* 2006. 45(8): p. 1198-1215.
[69] Schellekens, H., Bioequivalence and the immunogenicity of biopharmaceuticals. *Nat. Rev. Drug Discov.* 2002. 1(6): p. 457-462.
[70] Nair, L.S. and C.T. Laurencin, Polymers as biomaterials for tissue engineering and controlled drug delivery. *Adv. Biochem. Eng. Biotechnol.* 2006. 102: p. 47-90.
[71] Schellekens, H., Immunogenicity of therapeutic proteins: clinical implications and future prospects. *Clin. Ther.* 2002. 24(11): p. 1720-1740; discussion. 1719.
[72] Devalapally, H., A. Chakilam, and M.M. Amiji, Role of nanotechnology in pharmaceutical product development. *J. Pharm. Sci.* 2007. 96(10): p. 2547-2565.
[73] Kim, S., et al., Engineered polymers for advanced drug delivery. *Eur. J. Pharm. Biopharm.* 2009. 71(3): p. 420-430.
[74] Lukyanov, A.N. and V.P. Torchilin, Micelles from lipid derivatives of water-soluble polymers as delivery systems for poorly soluble drugs. *Adv. Drug Deliv. Rev.* 2004. 56(9): p. 1273-1289.
[75] Branco, M.C. and J.P. Schneider, Self-assembling materials for therapeutic delivery. *Acta Biomater.* 2009. 5(3): p. 817-831.
[76] Nayak, S. and L.A. Lyon, Soft nanotechnology with soft nanoparticles. *Angew. Chem. Int. Ed. Engl.* 2005. 44(47): p. 7686-7708.

[77] Murthy, N., et al., A novel strategy for encapsulation and release of proteins: hydrogels and microgels with acid-labile acetal cross-linkers. *J. Am. Chem. Soc.* 2002. 124(42): p. 12398-12399.

[78] Murthy, N., et al., A macromolecular delivery vehicle for protein-based vaccines: acid-degradable protein-loaded microgels. *Proc. Natl. Acad. Sci. U. S. A.* 2003. 100(9): p. 4995-5000.

[79] Leonard, M., et al., Hydrophobically modified alginate hydrogels as protein carriers with specific controlled release properties. *J. Control Release.* 2004. 98(3): p. 395-405.

[80] Giudice, E.L. and J.D. Campbell, Needle-free vaccine delivery. *Adv. Drug Deliv. Rev.* 2006. 58(1): p. 68-89.

[81] Alpar, H.O., et al., Biodegradable mucoadhesive particulates for nasal and pulmonary antigen and DNA delivery. *Adv. Drug Deliv. Rev.* 2005. 57(3): p. 411-430.

[82] Soane, R.J., et al., Evaluation of the clearance characteristics of bioadhesive systems in humans. *Int. J. Pharm.* 1999. 178(1): p. 55-65.

[83] Illum, L., Nasal drug delivery--possibilities, problems and solutions. *J. Control Release.* 2003. 87(1-3): p. 187-198.

[84] Norris, D.A., N. Puri, and P.J. Sinko, The effect of physical barriers and properties on the oral absorption of particulates. *Adv. Drug Deliv. Rev.* 1998. 34(2-3): p. 135-154.

[85] Lee, Y., et al., Conjugation of low-molecular-weight heparin and deoxycholic acid for the development of a new oral anticoagulant agent. *Circulation.* 2001. 104(25): p. 3116-3120.

[86] Lin, Y.H., et al., Multi-ion-crosslinked nanoparticles with pH-responsive characteristics for oral delivery of protein drugs. *J. Control Release.* 2008. 132(2): p. 141-149.

[87] Mi, F.L., et al., Oral delivery of peptide drugs using nanoparticles self-assembled by poly(gamma-glutamic acid) and a chitosan derivative functionalized by trimethylation. *Bioconjug Chem.* 2008. 19(6): p. 1248-1255.

[88] Thanou, M., J.C. Verhoef, and H.E. Junginger, Oral drug absorption enhancement by chitosan and its derivatives. *Adv. Drug Deliv. Rev.* 2001. 52(2): p. 117-126.

[89] Bravo-Osuna, I., et al., Mucoadhesion mechanism of chitosan and thiolated chitosan-poly(isobutyl cyanoacrylate) core-shell nanoparticles. *Biomaterials.* 2007. 28(13): p. 2233-2243.

[90] Lin, Y., Chen, CT., Liang, HF., Kulkarni, A. R., Lee, PW., Chen, CH., Sung, HW, Novel nanoparticles for oral insulin delivery via the paracellular pathway. *Nanotechnology*. 2007. 18: p. 105102-105113.
[91] Sonaje, K., et al., In vivo evaluation of safety and efficacy of self-assembled nanoparticles for oral insulin delivery. *Biomaterials*. 2009. 30(12): p. 2329-2339.
[92] Kotze, A.F., et al., Comparison of the effect of different chitosan salts and N-trimethyl chitosan chloride on the permeability of intestinal epithelial cells (Caco-2). *J. Control Release*. 1998. 51(1): p. 35-46.
[93] Qian, F., et al., Chitosan graft copolymer nanoparticles for oral protein drug delivery: preparation and characterization. *Biomacromolecules*. 2006. 7(10): p. 2722-2727.
[94] Jintapattanakit, A., et al., Peroral delivery of insulin using chitosan derivatives: a comparative study of polyelectrolyte nanocomplexes and nanoparticles. *Int. J. Pharm.* 2007. 342(1-2): p. 240-249.
[95] Mi, F.-L., Wu, Y-Y., Lin, Y-H., Sonaje, K., Ho, Y-C., Chen, C-T., Juang, J-H., Sung, H-W, Oral Delivery of Peptide Drugs Using Nanoparticles Self-Assembled by Poly(γ-glutamic acid) and a Chitosan Derivative Functionalized by Trimethylation. *Bioconjugate Chem*. 2008. 19(2): p. 1248-1255.
[96] Yin, L., et al., Drug permeability and mucoadhesion properties of thiolated trimethyl chitosan nanoparticles in oral insulin delivery. *Biomaterials*. 2009. 30(29): p. 5691-5700.
[97] Soane, R.J., et al., Clearance characteristics of chitosan based formulations in the sheep nasal cavity. *Int. J. Pharm.* 2001. 217(1-2): p. 183-191.
[98] Davis, S.S., Nasal vaccines. *Adv. Drug Deliv. Rev.* 2001. 51(1-3): p. 21-42.
[99] Illum, L., N.F. Farraj, and S.S. Davis, Chitosan as a novel nasal delivery system for peptide drugs. *Pharm. Res.* 1994. 11(8): p. 1186-1189.
[100] Illum, L., et al., Chitosan as a novel nasal delivery system for vaccines. *Adv. Drug Deliv. Rev.* 2001. 51(1-3): p. 81-96.
[101] Almeida, A.J. and H.O. Alpar, Nasal delivery of vaccines. *J. Drug Target*. 1996. 3(6): p. 455-467.
[102] Amidi, M., et al., Preparation and characterization of protein-loaded N-trimethyl chitosan nanoparticles as nasal delivery system. *J. Control Release,* 2006. 111(1-2): p. 107-116.

[103] Zhang, X., et al., Nasal absorption enhancement of insulin using PEG-grafted chitosan nanoparticles. *Eur. J. Pharm. Biopharm.* 2008. 68(3): p. 526-534.

[104] Wang, X., et al., Chitosan-NAC Nanoparticles as a Vehicle for Nasal Absorption Enhancement of Insulin. *J. Biomed. Mater. Res. Part B: Appl. Biomater. B,* 2009. 88: p. 150-161.

[105] Miller, C.J., M. McChesney, and P.F. Moore, Langerhans cells, macrophages and lymphocyte subsets in the cervix and vagina of rhesus macaques. *Lab. Invest.* 1992. 67(5): p. 628-634.

[106] Westerink, M.A., et al., ProJuvant (Pluronic F127/chitosan) enhances the immune response to intranasally administered tetanus toxoid. *Vaccine.* 2001. 20(5-6): p. 711-723.

[107] Kohler, D., Aerosols for systemic treatment. *Lung.* 1990. 168 Suppl: p. 677-684.

[108] Yamamoto, A., [Improvement of transmucosal absorption of biologically active peptide drugs]. *Yakugaku Zasshi.* 2001. 121(12): p. 929-948.

[109] Garcia-Contreras, L., et al., Evaluation of novel particles as pulmonary delivery systems for insulin in rats. *AAPS Pharm. Sci.* 2003. 5(2): article 9. DOI: 10.1208/ps050209

[110] Bosquillon, C., et al., Influence of formulation excipients and physical characteristics of inhalation dry powders on their aerosolization performance. *J. Control Release.* 2001. 70(3): p. 329-339.

[111] Courrier, H.M., N. Butz, and T.F. Vandamme, Pulmonary drug delivery systems: recent developments and prospects. *Crit. Rev. Ther. Drug Carrier Syst.* 2002. 19(4-5): p. 425-498.

[112] Ahsan, F., et al., Targeting to macrophages: role of physicochemical properties of particulate carriers--liposomes and microspheres--on the phagocytosis by macrophages. *J. Control Release.* 2002. 79(1-3): p. 29-40.

[113] Duszyk, M., CFTR and lysozyme secretion in human airway epithelial cells. *Pflugers Arch.* 2001. 443 Suppl 1: p. S45-49.

[114] Grenha, A., B. Seijo, and C. Remunan-Lopez, Microencapsulated chitosan nanoparticles for lung protein delivery. *Eur. J. Pharm. Sci.* 2005. 25(4-5): p. 427-437.

[115] Asghar, L.F. and S. Chandran, Multiparticulate formulation approach to colon specific drug delivery: current perspectives. *J. Pharm. Pharm. Sci.* 2006. 9(3): p. 327-338.

[116] Bayat, A., et al., Nanoparticles of quaternized chitosan derivatives as a carrier for colon delivery of insulin: ex vivo and in vivo studies. *Int. J. Pharm.* 2008. 356(1-2): p. 259-266.
[117] Pan, Y., et al., Bioadhesive polysaccharide in protein delivery system: chitosan nanoparticles improve the intestinal absorption of insulin in vivo. *Int. J. Pharm.* 2002. 249(1-2): p. 139-147.
[118] Huang, X., et al., Preparation and pharmacodynamics of low-molecular-weight chitosan nanoparticles containing insulin. *Carbohydrate Polymers.* 2009. 76(3): p. 368-373.
[119] Prego, C., et al., Chitosan-PEG nanocapsules as new carriers for oral peptide delivery. Effect of chitosan pegylation degree. *J. Control Release.* 2006. 111(3): p. 299-308.
[120] Yamamoto, H., et al., Surface-modified PLGA nanosphere with chitosan improved pulmonary delivery of calcitonin by mucoadhesion and opening of the intercellular tight junctions. *J. Control Release.* 2005. 102(2): p. 373-381.
[121] Chen, M.C., et al., The characteristics, biodistribution and bioavailability of a chitosan-based nanoparticulate system for the oral delivery of heparin. *Biomaterials.* 2009. 30(34): p. 6629-6637.
[122] Oyarzun-Ampuero, F.A., et al., Chitosan-hyaluronic acid nanoparticles loaded with heparin for the treatment of asthma. *Int. J. Pharm.* 2009. 381(2): p. 122-129.
[123] De Campos, A.M., A. Sanchez, and M.J. Alonso, Chitosan nanoparticles: a new vehicle for the improvement of the delivery of drugs to the ocular surface. Application to cyclosporin A. *Int. J. Pharm.* 2001. 224(1-2): p. 159-168.
[124] Colonna, C., et al., Ex vivo evaluation of prolidase loaded chitosan nanoparticles for the enzyme replacement therapy. *Eur. J. Pharm. Biopharm.* 2008. 70(1): p. 58-65.
[125] Colonna, C., et al., Site-directed PEGylation as successful approach to improve the enzyme replacement in the case of prolidase. *Int. J. Pharm.* 2008. 358(1-2): p. 230-237.
[126] Kim, J.H., et al., Self-assembled glycol chitosan nanoparticles for the sustained and prolonged delivery of antiangiogenic small peptide drugs in cancer therapy. *Biomaterials.* 2008. 29(12): p. 1920-1930.
[127] Vila, A., et al., Design of biodegradable particles for protein delivery. *J. Control Release.* 2002. 78(1-3): p. 15-24.

[128] Wang, S., et al., Preparation and evaluation of anti-neuroexcitation peptide (ANEP) loaded N-trimethyl chitosan chloride nanoparticles for brain-targeting. *Int. J. Pharm.* 2010. 386(1-2): p. 249-255.

[129] Mao, S., W. Sun, and T. Kissel, Chitosan-based formulations for delivery of DNA and siRNA. *Adv. Drug Deliv. Rev.* 2010. 62(1): p. 12-27.

[130] Hejazi, R. and M. Amiji, Chitosan-based gastrointestinal delivery systems. *J. Control Release.* 2003. 89(2): p. 151-165.

[131] Kim, T.H., et al., Chemical modification of chitosan as a gene carrier in vitro and in vivo. *Progress in Polymer Science.* 2007. 32(7): p. 726-753.

[132] Messai, I., et al., Poly(D,L-lactic acid) and chitosan complexes: interactions with plasmid DNA. *Colloids and Surfaces a-Physicochemical and Engineering Aspects.* 2005. 255(1-3): p. 65-72.

[133] Dai, H., et al., Chitosan-DNA nanoparticles delivered by intrabiliary infusion enhance liver-targeted gene delivery. *Int. J. Nanomedicine.* 2006. 1(4): p. 507-522.

[134] Strand, S.P., et al., Molecular design of chitosan gene delivery systems with an optimized balance between polyplex stability and polyplex unpacking. Biomaterials, 2010. 31(5): p. 975-987.

[135] Duceppe, N. and M. Tabrizian, Factors influencing the transfection efficiency of ultra low molecular weight chitosan/hyaluronic acid nanoparticles. *Biomaterials.* 2009. 30(13): p. 2625-2631.

[136] Huang, M., et al., Transfection efficiency of chitosan vectors: effect of polymer molecular weight and degree of deacetylation. *J. Control Release.* 2005. 106(3): p. 391-406.

[137] Koping-Hoggard, M., et al., Improved chitosan-mediated gene delivery based on easily dissociated chitosan polyplexes of highly defined chitosan oligomers. *Gene Ther.* 2004. 11(19): p. 1441-1452.

[138] Koping-Hoggard, M., et al., Chitosan as a nonviral gene delivery system. Structure-property relationships and characteristics compared with polyethylenimine in vitro and after lung administration in vivo. *Gene Ther.* 2001. 8(14): p. 1108-1121.

[139] Kiang, T., et al., The effect of the degree of chitosan deacetylation on the efficiency of gene transfection. *Biomaterials.* 2004. 25(22): p. 5293-5301.

[140] Lavertu, M., et al., High efficiency gene transfer using chitosan/DNA nanoparticles with specific combinations of molecular weight and degree of deacetylation. *Biomaterials.* 2006. 27(27): p. 4815-4824.

[141] Nafee, N., et al., Chitosan-coated PLGA nanoparticles for DNA/RNA delivery: effect of the formulation parameters on complexation and transfection of antisense oligonucleotides. *Nanomedicine.* 2007. 3(3): p. 173-183.
[142] Ishii, T., Y. Okahata, and T. Sato, Mechanism of cell transfection with plasmid/chitosan complexes. *Biochim. Biophys. Acta.* 2001. 1514(1): p. 51-64.
[143] Kim, T.H., et al., Galactosylated chitosan/DNA nanoparticles prepared using water-soluble chitosan as a gene carrier. *Biomaterials.* 2004. 25(17): p. 3783-3792.
[144] Douglas, K.L., C.A. Piccirillo, and M. Tabrizian, Effects of alginate inclusion on the vector properties of chitosan-based nanoparticles. *J. Control Release.* 2006. 115(3): p. 354-361.
[145] Peng, S.F., et al., Effects of incorporation of poly(gamma-glutamic acid) in chitosan/DNA complex nanoparticles on cellular uptake and transfection efficiency. *Biomaterials.* 2009. 30(9): p. 1797-1808.
[146] Germershaus, O., et al., Gene delivery using chitosan, trimethyl chitosan or polyethylenglycol-graft-trimethyl chitosan block copolymers: establishment of structure-activity relationships in vitro. *J. Control Release.* 2008. 125(2): p. 145-154.
[147] Kadiyala, I., et al., Transport of chitosan-DNA nanoparticles in human intestinal M-cell model versus normal intestinal enterocytes. *Eur. J. Pharm. Sci.* 2010. 39(1-3): p. 103-109.
[148] Sato, T., T. Ishii, and Y. Okahata, In vitro gene delivery mediated by chitosan. Effect of pH, serum, and molecular mass of chitosan on the transfection efficiency. *Biomaterials.* 2001. 22(15): p. 2075-2080.
[149] Jeong, J.H., S.W. Kim, and T.G. Park, Molecular design of functional polymers for gene therapy. *Progress in Polymer Science.* 2007. 32(11): p. 1239-1274.
[150] Jiang, H.L., et al., Galactosylated poly(ethylene glycol)-chitosan-graft-polyethylenimine as a gene carrier for hepatocyte-targeting. *J. Control Release.* 2008. 131(2): p. 150-157.
[151] Park, I.K., et al., Galactosylated chitosan (GC)-graft-poly(vinyl pyrrolidone) (PVP) as hepatocyte-targeting DNA carrier. Preparation and physicochemical characterization of GC-graft-PVP/DNA complex (1). *J. Control Release.* 2003. 86(2-3): p. 349-359.
[152] Dang, J.M. and K.W. Leong, Natural polymers for gene delivery and tissue engineering. *Adv. Drug Deliv. Rev.* 2006. 58(4): p. 487-499.

[153] Zheng, Y., et al., Receptor mediated gene delivery by folate conjugated N-trimethyl chitosan in vitro. *Int. J. Pharm.* 2009. 382(1-2): p. 262-269.
[154] Mao, H.Q., et al., Chitosan-DNA nanoparticles as gene carriers: synthesis, characterization and transfection efficiency. *J. Control Release.* 2001. 70(3): p. 399-421.
[155] de la Fuente, M., B. Seijo, and M.J. Alonso, Novel hyaluronic acid-chitosan nanoparticles for ocular gene therapy. *Invest. Ophthalmol. Vis. Sci.* 2008. 49(5): p. 2016-2024.
[156] Jiang, H.L., et al., Mannosylated chitosan-graft-polyethylenimine as a gene carrier for Raw 264.7 cell targeting. *Int. J. Pharm.* 2009. 375(1-2): p. 133-139.
[157] Lee, D., et al., Thiolated chitosan/DNA nanocomplexes exhibit enhanced and sustained gene delivery. *Pharm. Res.* 2007. 24(1): p. 157-167.
[158] Gao, Y., et al., Synthesis of 6-N,N,N-trimethyltriazole chitosan via "click chemistry" and evaluation for gene delivery. *Biomacromolecules.* 2009. 10(8): p. 2175-2182.
[159] Xu, Z.G., et al., Synthesis of biodegradable polycationic methoxy poly(ethylene glycol)-polyethylenimine-chitosan and its potential as gene carrier. *Carbohydrate Polymers.* 2009. 78(1): p. 46-53.
[160] Kim, T.H., et al., Synergistic effect of poly(ethylenimine) on the transfection efficiency of galactosylated chitosan/DNA complexes. *J. Control Release.* 2005. 105(3): p. 354-366.
[161] Jiang, H.L., et al., Chitosan-graft-polyethylenimine as a gene carrier. *J. Control Release.* 2007. 117(2): p. 273-280.
[162] Kim, T.H., et al., Efficient gene delivery by urocanic acid-modified chitosan. *J. Control Release.* 2003. 93(3): p. 389-402.
[163] Kiang, T., et al., Formulation of chitosan-DNA nanoparticles with poly(propyl acrylic acid) enhances gene expression. *J. Biomater. Sci. Polym. Ed.* 2004. 15(11): p. 1405-1421.
[164] Lee, J.I., K.S. Ha, and H.S. Yoo, Quantum-dot-assisted fluorescence resonance energy transfer approach for intracellular trafficking of chitosan/DNA complex. *Acta Biomater.* 2008. 4(4): p. 791-798.
[165] Hashimoto, M., et al., Lactosylated chitosan for DNA delivery into hepatocytes: the effect of lactosylation on the physicochemical properties and intracellular trafficking of pDNA/chitosan complexes. *Bioconjug. Chem.* 2006. 17(2): p. 309-316.

[166] Kim, T.H., et al., Mannosylated chitosan nanoparticle-based cytokine gene therapy suppressed cancer growth in BALB/c mice bearing CT-26 carcinoma cells. *Mol. Cancer Ther.* 2006. 5(7): p. 1723-1732.

[167] Fernandes, J.C., et al., Bone-protective effects of nonviral gene therapy with folate-chitosan DNA nanoparticle containing interleukin-1 receptor antagonist gene in rats with adjuvant-induced arthritis. *Mol. Ther.* 2008. 16(7): p. 1243-1251.

[168] Yuan, X., et al., Intranasal immunization with chitosan/pCETP nanoparticles inhibits atherosclerosis in a rabbit model of atherosclerosis. *Vaccine.* 2008. 26(29-30): p. 3727-3734.

[169] Kumar, M., et al., Chitosan IFN-gamma-pDNA Nanoparticle (CIN) Therapy for Allergic Asthma. *Genet. Vaccines Ther.* 2003. 1(1): p. 3. doi:10.1186/1479-0556-1-3.

[170] Bivas-Benita, M., et al., Pulmonary delivery of chitosan-DNA nanoparticles enhances the immunogenicity of a DNA vaccine encoding HLA-A*0201-restricted T-cell epitopes of Mycobacterium tuberculosis. *Vaccine.* 2004. 22(13-14): p. 1609-1615.

[171] Bowman, K., et al., Gene transfer to hemophilia A mice via oral delivery of FVIII-chitosan nanoparticles. *J. Control Release.* 2008. 132(3): p. 252-259.

[172] Khatri, K., et al., Plasmid DNA loaded chitosan nanoparticles for nasal mucosal immunization against hepatitis B. *Int. J. Pharm.* 2008. 354(1-2): p. 235-241.

[173] Xu, W., et al., Intranasal delivery of chitosan-DNA vaccine generates mucosal SIgA and anti-CVB3 protection. *Vaccine.* 2004. 22(27-28): p. 3603-3612.

[174] Iqbal, M., et al., Nasal delivery of chitosan-DNA plasmid expressing epitopes of respiratory syncytial virus (RSV) induces protective CTL responses in BALB/c mice. *Vaccine.* 2003. 21(13-14): p. 1478-1485.

[175] Liu, X., et al., The influence of polymeric properties on chitosan/siRNA nanoparticle formulation and gene silencing. *Biomaterials.* 2007. 28(6): p. 1280-1288.

[176] Howard, K.A., Delivery of RNA interference therapeutics using polycation-based nanoparticles. *Adv. Drug Deliv. Rev.* 2009. 61(9): p. 710-720.

[177] Lai, W.F. and M.C. Lin, Nucleic acid delivery with chitosan and its derivatives. *J. Control Release.* 2009. 134(3): p. 158-168.

[178] Katas, H. and H.O. Alpar, Development and characterization of chitosan nanoparticles for siRNA delivery. *J. Control Release.* 2006. 115(2): p. 216-225.
[179] Howard, K.A., et al., RNA interference in vitro and in vivo using a novel chitosan/siRNA nanoparticle system. *Mol. Ther.* 2006. 14(4): p. 476-484.
[180] Rojanarata, T., et al., Chitosan-thiamine pyrophosphate as a novel carrier for siRNA delivery. *Pharm. Res.* 2008. 25(12): p. 2807-2814.
[181] Lee, D.W., et al., Preparation and characterization of chitosan/polyguluronate nanoparticles for siRNA delivery. *J. Control Release.* 2009. 139(2): p. 146-152.
[182] Ji, A.M., et al., Functional gene silencing mediated by chitosan/siRNA nanocomplexes. *Nanotechnology.* 2009. 20(40): p. 405103. doi: 10.1088/0957-4484/20/40/405103 .
[183] Tahara, K., et al., Chitosan-modified poly(d,l-lactide-co-glycolide) nanospheres for improving siRNA delivery and gene-silencing effects. *Eur. J. Pharm. Biopharm.* 2010. 74(3): p. 421-426.
[184] Kong, X., et al., Respiratory syncytial virus infection in Fischer 344 rats is attenuated by short interfering RNA against the RSV-NS1 gene. *Genet. Vaccines Ther.* 2007. 5: p. 4. doi: 10.1186/1479-0556-5-4.
[185] Wang, X., et al., Prevention of airway inflammation with topical cream containing imiquimod and small interfering RNA for natriuretic peptide receptor. *Genet. Vaccines Ther.* 2008. 6: p. 7. doi:10.1186/1479-0556-6-7.
[186] Howard, K.A., et al., Chitosan/siRNA nanoparticle-mediated TNF-alpha knockdown in peritoneal macrophages for anti-inflammatory treatment in a murine arthritis model. *Mol. Ther.* 2009. 17(1): p. 162-168.
[187] Jere, D., et al., Chitosan-graft-polyethylenimine for Akt1 siRNA delivery to lung cancer cells. *Int. J. Pharm.* 2009. 378(1-2): p. 194-200.
[188] Liu, Z.H., et al., Polysaccharides-based nanoparticles as drug delivery systems. *Advanced Drug Delivery Reviews.* 2008. 60(15): p. 1650-1662.
[189] Agnihotri, S.A., N.N. Mallikarjuna, and T.M. Aminabhavi, Recent advances on chitosan-based micro- and nanoparticles in drug delivery. *Journal of Controlled Release.* 2004. 100(1): p. 5-28.
[190] Hafner, A., et al., Melatonin-loaded lecithin/chitosan nanoparticles: physicochemical characterisation and permeability through Caco-2 cell monolayers. *Int. J. Pharm.* 2009. 381(2): p. 205-213.
[191] Park, J.H., et al., Targeted delivery of low molecular drugs using chitosan and its derivatives. *Adv. Drug Deliv. Rev.* 2010. 62(1): p. 28-41.

[192] Bisht, S. and A. Maitra, Dextran-doxorubicin/chitosan nanoparticles for solid tumor therapy. *Wiley Interdiscip. Rev. Nanomed. Nanobiotechnol.* 2009. 1(4): p. 415-425.

[193] Kim, K., et al., Self-Assembled Nanoparticles of Bile Acid Modified Glycol Chitosans and Their Applications for Cancer Therapy. *Macromolecular Research.* 2005. 13(3): p. 167-175.

[194] Kim, J.H., et al., Hydrophobically modified glycol chitosan nanoparticles as carriers for paclitaxel. *J. Control Release.* 2006. 111(1-2): p. 228-234.

[195] Liu, K.H., et al., Self-assembly behavior and doxorubicin-loading capacity of acylated carboxymethyl chitosans. *J. Phys. Chem. B,* 2009. 113(35): p. 11800-11807.

[196] Lee, E., et al., Conjugated chitosan as a novel platform for oral delivery of paclitaxel. *J. Med. Chem.* 2008. 51(20): p. 6442-6449.

[197] Son, Y.J., et al., Biodistribution and anti-tumor efficacy of doxorubicin loaded glycol-chitosan nanoaggregates by EPR effect. *J. Control Release.* 2003. 91(1-2): p. 135-145.

[198] Hu, F.Q., et al., Shell cross-linked stearic acid grafted chitosan oligosaccharide self-aggregated micelles for controlled release of paclitaxel. *Colloids Surf. B Biointerfaces.* 2006. 50(2): p. 97-103.

[199] Hu, F.Q., et al., Cellular uptake and cytotoxicity of shell crosslinked stearic acid-grafted chitosan oligosaccharide micelles encapsulating doxorubicin. *Eur. J. Pharm. Biopharm.* 2008. 69(1): p. 117-125.

[200] Min, K.H., et al., Hydrophobically modified glycol chitosan nanoparticles-encapsulated camptothecin enhance the drug stability and tumor targeting in cancer therapy. *J. Control Release.* 2008. 127(3): p. 208-218.

[201] Fan, L., et al., Novel super pH-sensitive nanoparticles responsive to tumor extracellular pH. *Carbohydrate Polymers.* 2008. 73(3): p. 390-400.

[202] Hu, Y., et al., Hollow chitosan/poly(acrylic acid) nanospheres as drug carriers. *Biomacromolecules.* 2007. 8(4): p. 1069-1076.

[203] Hyung Park, J., et al., Self-assembled nanoparticles based on glycol chitosan bearing hydrophobic moieties as carriers for doxorubicin: in vivo biodistribution and anti-tumor activity. *Biomaterials.* 2006. 27(1): p. 119-126.

[204] Zhao, Z., et al., Biodegradable Nanoparticles Based on Linoleic Acid and Poly(beta-malic acid) Double Grafted Chitosan Derivatives as

Carriers of Anticancer Drugs. *Biomacromolecules.* 2009. 10(3): p. 565-572.

[205] Saravanakumar, G., et al., Hydrotropic oligomer-conjugated glycol chitosan as a carrier of paclitaxel: synthesis, characterization, and in vivo biodistribution. *J. Control Release.* 2009. 140(3): p. 210-217.

[206] Kim, J.H., et al., Antitumor efficacy of cisplatin-loaded glycol chitosan nanoparticles in tumor-bearing mice. *J. Control Release.* 2008. 127(1): p. 41-49.

[207] Bilensoy, E., et al., Intravesical cationic nanoparticles of chitosan and polycaprolactone for the delivery of Mitomycin C to bladder tumors. *Int. J. Pharm.* 2009. 371(1-2): p. 170-176.

[208] Zhang, D.Y., et al., Preparation of chitosan-polyaspartic acid-5-fluorouracil nanoparticles and its anti-carcinoma effect on tumor growth in nude mice. *World J. Gastroenterol.* 2008. 14(22): p. 3554-3562.

[209] Wang, Q., et al., Norcantharidin-associated galactosylated chitosan nanoparticles for hepatocyte-targeted delivery. *Nanomedicine.* 2010. doi:10.1016/j.nano.2009.07.006.

[210] Lee, S.J., et al., Tumor specificity and therapeutic efficacy of photosensitizer-encapsulated glycol chitosan-based nanoparticles in tumor-bearing mice. *Biomaterials.* 2009. 30(15): p. 2929-2939.

[211] Jeong, Y.I., et al., Polyion complex micelles composed of all-trans retinoic acid and poly (ethylene glycol)-grafted-chitosan. *J. Pharm. Sci.* 2006. 95(11): p. 2348-2360.

[212] Hu, F.Q., et al., PEGylated chitosan-based polymer micelle as an intracellular delivery carrier for anti-tumor targeting therapy. *Eur. J. Pharm. Biopharm.* 2008. 70(3): p. 749-757.

[213] Qu, G., et al., PEG conjugated N-octyl-O-sulfate chitosan micelles for delivery of paclitaxel: in vitro characterization and in vivo evaluation. *Eur. J. Pharm. Sci.* 2009. 37(2): p. 98-105.

[214] Park, J.S., et al., N-acetyl histidine-conjugated glycol chitosan self-assembled nanoparticles for intracytoplasmic delivery of drugs: endocytosis, exocytosis and drug release. *J. Control Release.* 2006. 115(1): p. 37-45.

[215] Mi, F.L., et al., Synthesis of a novel glycoconjugated chitosan and preparation of its derived nanoparticles for targeting HepG2 cells. *Biomacromolecules.* 2007. 8(3): p. 892-898.

[216] Dufes, C., et al., Anticancer drug delivery with transferrin targeted polymeric chitosan vesicles. *Pharm. Res.* 2004. 21(1): p. 101-107.

[217] You, J., et al., Folate-conjugated polymer micelles for active targeting to cancer cells: preparation, in vitro evaluation of targeting ability and cytotoxicity. *Nanotechnology.* 2008. 19(4): p. 045102. doi: 10.1088/0957-4484/19/04/045102.
[218] Jain, A. and S.K. Jain, In vitro and cell uptake studies for targeting of ligand anchored nanoparticles for colon tumors. *Eur. J. Pharm. Sci.* 2008. 35(5): p. 404-416.
[219] Nagarwal, R.C., et al., Polymeric nanoparticulate system: a potential approach for ocular drug delivery. *J. Control Release.* 2009. 136(1): p. 2-13.
[220] de la Fuente, M., et al., Chitosan-based Nanostructures: A Delivery Platform for Ocular Therapeutics. *Adv. Drug Deliv. Rev.* 2010. 62(1): p. 100-117.
[221] De Campos, A.M., et al., The effect of a PEG versus a chitosan coating on the interaction of drug colloidal carriers with the ocular mucosa. *Eur. J. Pharm. Sci.* 2003. 20(1): p. 73-81.
[222] de Campos, A.M., et al., Chitosan nanoparticles as new ocular drug delivery systems: in vitro stability, in vivo fate, and cellular toxicity. *Pharm. Res.* 2004. 21(5): p. 803-810.
[223] Motwani, S.K., et al., Chitosan-sodium alginate nanoparticles as submicroscopic reservoirs for ocular delivery: formulation, optimisation and in vitro characterisation. *Eur. J. Pharm. Biopharm.* 2008. 68(3): p. 513-525.
[224] Badawi, A.A., et al., Chitosan based nanocarriers for indomethacin ocular delivery. *Arch. Pharm. Res.* 2008. 31(8): p. 1040-1049.
[225] Yang, K.W., et al., Novel polyion complex micelles for liver-targeted delivery of diammonium glycyrrhizinate: in vitro and in vivo characterization. *J. Biomed. Mater. Res. A,* 2009. 88(1): p. 140-148.
[226] Lin, A., et al., Glycyrrhizin surface-modified chitosan nanoparticles for hepatocyte-targeted delivery. *Int. J. Pharm.* 2008. 359(1-2): p. 247-253.
[227] Lin, A., et al., Preparation and evaluation of N-caproyl chitosan nanoparticles surface modified with glycyrrhizin for hepatocyte targeting. *Drug Dev. Ind. Pharm.* 2009. 35(11): p. 1348-1355.
[228] Wang, X., N. Chi, and X. Tang, Preparation of estradiol chitosan nanoparticles for improving nasal absorption and brain targeting. *Eur. J. Pharm. Biopharm.* 2008. 70(3): p. 735-740.
[229] Wang, Z.H., et al., Trimethylated chitosan-conjugated PLGA nanoparticles for the delivery of drugs to the brain. *Biomaterials.* 31(5): p. 908-915.

[230] Azarmi, S., W.H. Roa, and R. Lobenberg, Targeted delivery of nanoparticles for the treatment of lung diseases. *Adv. Drug Deliv. Rev.* 2008. 60(8): p. 863-875.

[231] Lee, D.W., et al., Thiolated chitosan nanoparticles enhance anti-inflammatory effects of intranasally delivered theophylline. *Respir. Res.* 2006. 7: p. 112. doi: 10.1186/1465-9921-7-112.

Index

A

absorption, 12, 14, 15, 16, 17, 18, 19, 20, 45, 55, 57, 59, 60, 68
acetic acid, 3
acid, 12, 13, 20, 21, 24, 27, 28, 29, 34, 39, 40, 41, 42, 51, 52, 53, 54, 56, 57, 60, 61, 63, 64, 66, 67
acrylic acid, 29, 39, 55, 63, 66
active transport, 44
additives, 23
adhesion, 43
adsorption, 28
aggregation, 24, 25, 33, 53
airway epithelial cells, 59
airway hyperresponsiveness, 36, 45
airway inflammation, 31, 65
airways, 17
albumin, 26
allergic asthma, 30, 45
alveolar macrophage, 17, 44
alveoli, 17
amines, 3, 22
angiogenesis, 31
antibiotic, 43, 52
anticancer drug, 37
anticoagulant, 20, 57
antigen, 57
anti-inflammatory drugs, 54
antisense, 22, 62
antisense oligonucleotides, 62
antitumor, 39, 40, 41, 56
apoptosis, 31, 40
arthritis, 31, 36, 64, 65
ascending colon, 19
asthma, 36, 45, 60
atherosclerosis, 30, 31, 64
ATP, 28
autoimmune diseases, 11

B

barriers, 25, 37, 43, 57
BBB, 44
bioavailability, 3, 11, 12, 15, 37, 43, 60
biocompatibility, 16, 22
biodegradability, 1, 22
biomaterials, 56
biomedical applications, 1, 2, 11
biopolymer, 24
blood stream, 20, 25, 38
bloodstream, 26
bonds, 15
bone, 2, 25, 31, 52
bone marrow, 26
bone resorption, 31

C

calcitonin, 20, 60
calcium, 20, 52
cancer, 11, 30, 35, 36, 37, 40, 41, 54, 60, 64, 66, 68
cancer cells, 37, 41, 68
cancerous cells, 42
carbohydrate, 42
carcinoma, 31, 34, 40, 64, 67
cartilage, 36
cell cycle, 31
cell line, 27, 28, 40, 42
cell membranes, 12, 41
cell surface, 12, 13, 26, 28, 41, 43
cellulose, 1
central nervous system, 44
cervix, 59
charge density, 23
chemical stability, 3
chitin, 1, 2, 52, 54
chronic diseases, 35
circulation, 11, 25, 26, 38, 39, 40, 41
clarity, 2
coenzyme, 44
collagen, 52
colon, 12, 18, 42, 53, 59, 60, 68
colorectal cancer, 35
complex interactions, 28
compliance, 12, 16
compounds, 2, 14
condensation, 24
conjugation, 27, 44
conjunctiva, 21, 43
consensus, 17
contact time, 12
copolymers, 62
cornea, 21, 43
cortex, 44
cost, 8, 16
covalent bond, 15, 28

brain, 21, 44, 61, 68
breakdown, 41
breast cancer, 39

covalent bonding, 15
coxsackievirus, 30
culture, 24
cytochrome, 18
cytokines, 31, 36
cytoplasm, 29, 30, 41
cytoskeleton, 14
cytotoxicity, 29, 39, 42, 56, 66, 68

D

defense mechanisms, 17
deficiency, 21
deformability, 40
degradation, 3, 22, 23, 25, 28, 29, 33, 37
denaturation, 12
dendritic cell, 31
deposition, 17
derivatives, 2, 18, 22, 27, 37, 47, 56, 57, 58, 60, 64, 65
destination, 30
destruction, 36
detergents, 37
diffusion, 8, 20
digestion, 34
digestive enzymes, 13
discomfort, 11
dispersion, 4
dissociation, 23, 24, 28
distress, 37
DNA, 6, 22, 23, 24, 25, 26, 27, 28, 29, 30, 31, 32, 33, 55, 57, 61, 62, 63, 64
DNase, 23
docetaxel, 56
dosing, 11
drug carriers, 2, 6, 38, 66
drug delivery, vii, 2, 3, 6, 11, 17, 18, 37, 43, 47, 51, 53, 54, 55, 56, 57, 58, 59, 65, 67, 68
drug release, 37, 39, 41, 44, 67
drug therapy, 44
drugs, 3, 4, 11, 17, 18, 29, 37, 38, 39, 41, 42, 43, 51, 56, 57, 58, 59, 60, 65, 67, 68
duodenum, 13

Index

E

emulsions, 4, 5
encapsulation, 35, 38, 57
encoding, 32, 64
endothelial cells, 44
endothelium, 38, 44
energy transfer, 63
engineering, 2, 52, 56, 62
entrapment, 4, 30
enzymatic activity, 17, 18
enzymes, 18
eosinophilia, 36
eosinophils, 45
epithelia, 16, 17
epithelial cells, 13, 58
epithelium, 12, 14, 17, 43
ester, 31
ethylene, 26, 27, 40, 41, 55, 62, 63, 67
ethylene glycol, 26, 27, 40, 41, 62, 63, 67
ethylene oxide, 55
evaporation, 20, 21
exocytosis, 67
exoskeleton, 1
exposure, 36
extravasation, 38

F

fabrication, 55
fasting, 13
fibroblasts, 21, 52
films, 7, 53
fluid, 17
fluorescence, 29, 63
folate, 27, 42, 55, 63, 64
free energy, 9

G

gadolinium, 54
gelation, 6, 13, 16, 20, 21, 34
gene expression, 22, 23, 27, 28, 29, 30, 35, 36, 63

gene silencing, 33, 35, 64, 65
gene therapy, 22, 31, 55, 62, 63, 64
gene transfer, 24, 28, 33, 61
genes, 22, 25, 33
genetic information, 30
glucose, 14, 19, 20
glutamate, 34
glutamic acid, 13, 24, 57, 58, 62
glycol, 16, 21, 39, 41, 42, 56, 60, 66, 67
glycoproteins, 15
glycosaminoglycans, 2

H

hemophilia, 30, 31, 64
hepatitis, 30, 43, 64
hepatocytes, 44, 63
hepatoma, 42
histamine, 20
histidine, 41, 67
HLA, 64
hydrogels, vii, 1, 11, 54, 57
hydrogen, 22
hydrolysis, 30
hydrophilicity, 12
hydroxyapatite, 52
hydroxyl, 1
hyperplasia, 45

I

IFN, 31, 64
ileum, 13
immune response, 17, 22, 32, 59
immunity, 17, 32
immunization, 16, 64
immunogenicity, 12, 56, 64
in vivo, 13, 22, 24, 26, 28, 32, 33, 35, 37, 42, 43, 47, 53, 60, 61, 65, 66, 67, 68
incubation time, 44
inflammation, 36
inflammatory disease, 45
inhibition, 35, 40
inhibitor, 30

insects, 1
insulin, 13, 14, 15, 16, 18, 20, 55, 58, 59, 60
interference, 33, 64, 65
internalization, 24, 26, 41, 42, 44
intestine, 13
intravenously, 21
ionic strength, 24
ionization, 24
ions, 16, 28
irradiation, 40
isozymes, 18

J

jejunum, 13
joints, 36

L

lactic acid, 21, 61
lactose, 18, 43
Langerhans cells, 59
lecithin, 65
lens, 2
lesions, 31
ligand, 27, 42, 44, 68
liposomes, 11, 22, 59
liver, 25, 27, 31, 39, 41, 42, 43, 61, 68
liver cancer, 42
liver cells, 27, 41
localization, 30, 41
luciferase, 23, 28, 30
lumen, 12, 14
lung cancer, 65
lung disease, 69
lymph, 43
lymphocytes, 32
lysosome, 30, 33
lysozyme, 18, 21, 59

M

macromolecules, 6, 12, 42
macrophages, 17, 25, 36, 41, 59, 65

mannitol, 18
mast cells, 20
matrix, 3, 13, 42
mechanical properties, 6, 52
melanoma, 53
membranes, vii, 1, 29
messenger RNA, 33
metabolic disorder, 11
metabolism, 16
methacrylic acid, 55
methanol, 4
methodology, 4
methyl methacrylate, 15
mice, 31, 32, 36, 39, 40, 44, 64, 67
micelles, 7, 8, 11, 40, 41, 42, 43, 55, 56, 66, 67, 68
microemulsion, 7
microgels, 57
micrometer, 18
microspheres, 3, 18, 51, 54, 59
microstructure, 52
migration, 40
Mitoxantrone, 54
MMA, 15
molecular mass, 62
molecular structure, 55
molecular weight, 1, 11, 37, 39, 61
molecules, 3, 7, 8, 12, 41
monomers, 7
morphology, 31
mRNA, 33
mucin, 15, 16, 28
mucosa, 13, 15, 16, 45, 68
mucus, 13, 14, 15, 17, 45

N

nanocomposites, 52
nanometer, 11
nanoparticles, 3, 53, 54, 55, 56, 57, 58, 59, 60, 61, 62, 63, 64, 65, 66, 67, 68, 69
nanotechnology, 56
nerve, 52
nitrogen, 23
nuclear magnetic resonance, 55

nucleic acid, 33, 34
nucleotides, 33
nucleus, 25, 30, 41, 42

O

obstacles, 16, 17
obstruction, 30
oil, 5, 7, 55
oligomers, 23, 61
oligosaccharide, 19, 20, 29, 41, 56, 66
organ, 37
organic solvents, 1, 4, 37

P

PAA, 39
paclitaxel, 41, 56, 66, 67
pancreas, 13
parenchyma, 44
pathways, 12, 27
peptidase, 18
peptides, 12, 17, 18, 19, 20
performance, 6, 42, 59
periodontal, 2
permeability, 2, 12, 13, 28, 37, 38, 39, 45, 58, 65
permeation, 14, 15, 34
phagocytosis, 18, 59
pharmacokinetics, 37
phosphates, 52
physicochemical properties, 3, 4, 59, 63
plasma proteins, 26
plasmid, 29, 61, 62, 64
platform, 66
polymer, 11, 21, 23, 28, 29, 30, 38, 61, 67, 68
polymer chains, 23
polymerization, 7
polymers, 15, 22, 26, 28, 29, 56, 62
prevention, 24, 44
probability, 18
probe, 44
pro-inflammatory, 36

proliferation, 52
properties, vii, 1, 6, 12, 14, 18, 25, 37, 45, 51, 53, 55, 56, 57, 58, 62, 64
propylene, 55
protease inhibitors, 18
proteins, 4, 11, 12, 13, 16, 17, 18, 19, 20, 26, 41, 47, 55, 56, 57
proteolysis, 11
protons, 28
pumps, 28
purification, 4
PVP, 26, 62
pyrophosphate, 34, 65

R

radical polymerization, 15
radius, 33, 34
reactions, 15
reactive groups, 1
reactivity, 1, 2
receptors, 26, 27, 41
recognition, 42
redistribution, 12, 13
replacement, 2, 21, 44, 60
requirements, 3, 25
residues, 1, 15, 34
respiratory syncytial virus, 30, 64
rheumatoid arthritis, 30
RNA, 33, 34, 62, 64, 65
RNAi, 33

S

salmon, 20
salt formation, 34
salts, 34, 58
secretion, 31, 52, 59
self-assembly, 8, 15
sensitivity, 13, 29
sensitization, 31
serum, 23, 24, 25, 28, 31, 32, 33, 47, 62
serum albumin, 25
sex, 44

sheep, 58
sialic acid, 15
side effects, 11, 37, 45
signals, 41
siRNA, 22, 33, 34, 35, 36, 61, 64, 65
skin, 2, 21, 52
small intestine, 13, 15, 18
sodium, 4, 34, 43, 54, 68
sodium hydroxide, 4
solid tumors, 42
solubility, 1, 14, 15, 24, 25, 34, 51
spatial memory, 44
species, 6
spleen, 25, 31, 43
sponge, 28, 29, 52
Sprague-Dawley rats, 15
stoichiometry, 23
stomach, 13
strategy, 9, 38, 41, 57
strong interaction, 24
structural modifications, 1
structural protein, 32
subdomains, 28
substitutes, 2
suppression, 39
surface area, 17
surfactant, 3, 7
survival, 26
susceptibility, 11
swelling, 28
synergistic effect, 17, 28
synthesis, 33, 63, 67
systemic immune response, 16, 21

T

tetanus, 17, 21, 59
therapeutic agents, 3
therapeutics, 11, 56, 64
therapy, 22, 35, 37, 38, 41, 53, 54, 60, 66, 67
tissue, 26, 38, 39, 41, 52, 54, 56, 62
TMC, 15, 16, 21, 24, 27, 44
TNF, 36, 65
TNF-alpha, 65

TNF-α, 36
toxicity, 1, 22, 39, 40, 53, 68
trachea, 16
transcription, 29, 30
transfection, 22, 23, 24, 26, 27, 28, 29, 34, 56, 61, 62, 63
transferrin, 27, 42, 67
translation, 24
translocation, 25, 29, 30
transport, 12, 14, 16, 19, 24, 25, 27, 30, 37, 44
triggers, 13
tuberculosis, 30, 31, 64
tumor, 21, 31, 38, 39, 40, 41, 66, 67
tumor cells, 31, 39, 42
tumor growth, 21, 31, 39, 67
tumors, 39, 42, 67, 68
turnover, 31

V

vaccine, 16, 32, 57, 64
vagina, 59
vasculature, 38
vector, 34, 62
vein, 44
ventricle, 44
viral vectors, 22
virus infection, 32, 65
viscosity, 16
vulnerability, 33

W

water-soluble polymers, 56
wettability, 2
wound healing, 2
wound infection, 52

X

xenografts, 39